Rights for Robots

Bringing a unique perspective to the burgeoning ethical and legal issues surrounding the presence of artificial intelligence in our daily lives, the book uses theory and practice on animal rights and the rights of nature to assess the status of robots.

Through extensive philosophical and legal analyses, the book explores how rights can be applied to nonhuman entities. This task is completed by developing a framework useful for determining the kinds of personhood for which a nonhuman entity might be eligible, and a critical environmental ethic that extends moral and legal consideration to nonhumans. The framework and ethic are then applied to two hypothetical situations involving real-world technology—animal-like robot companions and humanoid sex robots. Additionally, the book approaches the subject from multiple perspectives, providing a comparative study of legal cases on animal rights and the rights of nature from around the world and insights from structured interviews with leading experts in the field of robotics. Ending with a call to rethink the concept of rights in the Anthropocene, suggestions for further research are made.

An essential read for scholars and students interested in robot, animal, and environmental law, as well as those interested in technology more generally, the book is a ground-breaking study of an increasingly relevant topic, as robots become ubiquitous in modern society.

Joshua C. Gellers is an associate professor of Political Science at the University of North Florida, Research Fellow of the Earth System Governance Project, and Core Team Member of the Global Network for Human Rights and the Environment. His research focuses on the relationship between the environment, human rights, and technology. Josh has published work in *Global Environmental Politics, International Environmental Agreements*, and *Journal of Environment and Development*, among others. He is the author of *The Global Emergence of Constitutional Environmental Rights* (Routledge 2017).

Rights for Robots

Artificial Intelligence, Animal and
Environmental Law

Joshua C. Gellers

 Routledge
Taylor & Francis Group

LONDON AND NEW YORK

First published 2021
by Routledge
2 Park Square, Milton Park, Abingdon, Oxon OX14 4RN

and by Routledge
52 Vanderbilt Avenue, New York, NY 10017

Routledge is an imprint of the Taylor & Francis Group, an informa business

British Library Cataloguing-in-Publication Data
A catalogue record for this book is available from the British Library

Library of Congress Cataloging-in-Publication Data
A catalog record has been requested for this book

ISBN: 9780367211745 (hbk)
ISBN: 9780429288159 (ebk)

Typeset in Times New Roman
by Deanta Global Publishing Services, Chennai, India

To Allie, my sunshine,

and Lillie Faye, our sky.

Contents

Figures

Tables

Acknowledgments

Many individuals assisted me in the completion of this book in ways big and small. I wish to recognize them here as a token of my gratitude. Thanks to my editor at Routledge, Colin Perrin, who saw promise in the project that later blossomed into this book. In the environmental domain, David Boyd, Erin Daly, Anna Grear, Craig Kauffman, David Vogel, and participants in the "Earth System Governance 4.0" panel at the 2019 Mexico Conference on Earth System Governance supplied key insights that helped me look at robots through an ecological lens. A number of experts on artificial intelligence (AI) and robotics welcomed this outsider into conversations about philosophical and legal issues surrounding intelligent machines. They include Joanna Bryson, Mark Coeckelbergh, Kate Devlin, Daniel Estrada, David Gunkel, Paresh Kathrani, and Noel Sharkey. I am humbled by those roboticists and technologists who were willing to be interviewed for this project. These gracious interviewees include Kate Darling, Yoshikazu Kanamiya, Takayuki Kanda, Ryutaro Murayama, Atsuo Takanishi, Fumihide Tanaka, Yueh-Hsuan Weng, and Jinseok Woo. I could not have conducted field research in Japan without the financial support of a faculty development grant from my institution, the University of North Florida (UNF). At UNF, I am grateful for Mandi Barringer, who let me talk about robots with her students, and Ayan Dutta, who took the time to discuss swarm robotics with me. Also, I am deeply appreciative of the work put in by Patrick Healy, who transcribed all of my interviews. Thanks to my mom, dad, and Aunt Diane, who encouraged my love of science and science fiction; my brother Brett and sister-in-law Jessica, who met my robotic musings with healthy skepticism; my late uncle Tuvia Ben-Shmuel Yosef (Don Gellers), whose advocacy for the Passamaquoddy Tribe in Maine has deservedly earned him posthumous justice and acclaim; my dog Shiva, who participated in several lay experiments that confirmed her possession of consciousness, intelligence, and intentionality; and my dear wife Allie, whose unyielding love for me might only be surpassed by the patience she has exhibited throughout this entire process. There's no one else with whom I'd rather be self-quarantined.

Abbreviations

Association of Professional Lawyers for Animal Rights	(AFADA)
Artificial Intelligence	(AI)
Community Environmental Legal Defense Fund	(CELDF)
Center for Great Apes	(CGA)
Human–Robot Interaction	(HRI)
Information Ethics	(IE)
International Organization for Standardization	(ISO)
Nationally Determined Contributions	(NDCs)
Nonhuman Rights Project	(NhRP)
Rights of Nature	(RoN)
Universal Declaration on Human Rights	(UDHR)
Unmanned Aerial Vehicles	(UAVs)

Introduction

Theodore: Cause you seem like a person, but you're just a voice in a computer.
Samantha: I can understand how the limited perspective of an un-artificial mind
would perceive it that way. You'll get used to it.[1]

Can robots have rights? This question has inspired significant debate among philosophers, computer scientists, policymakers, and the popular press. However, much of the discussion surrounding this issue has been conducted in the limited quarters of disciplinary silos and without a fuller appreciation of important macro-level developments. I argue that the so-called "machine question" (Gunkel, 2012, p. x), specifically the inquiry into whether and to what extent intelligent machines might warrant moral (or perhaps legal) consideration, deserves extended analysis in light of these developments.

Two global trends seem to be on a collision course. On the one hand, robots are becoming increasingly human-like in their appearance and behavior. Sophia, a female-looking humanoid robot created by Hong Kong–based Hanson Robotics (*Hi, I Am Sophia...*, 2019), serves as a prime example. In 2017, Sophia captured the world's imagination (and drew substantial ire as well) when the robot was granted "a citizenship" by the Kingdom of Saudi Arabia (Hatmaker, 2017). While this move was criticized as a "careful piece of marketing" (British Council, n.d.), "eroding human rights" (Hart, 2018), and "obviously bullshit" (J. Bryson quoted in Vincent, 2017), it elevated the idea that robots might be eligible for certain types of legal status based on how they look and act. Despite the controversy surrounding Sophia and calls to temper the quest for human-like appearance, the degree to which robots are designed to emulate humans is only likely to increase in the future, be it for reasons related to improved functioning in social environments or the hubris of roboticists.

On the other hand, legal systems around the world are increasingly recognizing the rights of nonhuman entities. The adoption of Ecuador's 2008 Constitution marked a watershed moment in this movement, as the charter devoted an entire chapter to the rights of nature (RoN) (Ecuador Const., tit. II, ch. 7). Courts and legislatures different corners of the globe have similarly identified rights held by nonhumans—the Whanganui River in New Zealand, the Ganges and its tributaries in India, the Atrato River in Colombia, and Mother Nature herself (*Pachamama*) in Ecuador (Cano-Pecharroman, 2018). In the United States, nearly 100 municipal ordinances invoking the RoN have been passed or pending since 2006 (Kauffman & Martin, 2018, p. 43). Many more efforts to legalize the RoN are afoot at the subnational, national, and international levels (Global Alliance for the Rights of Nature, 2019). All of this is happening in tandem with legal efforts seeking to protect animals under the argument that they, too, possess rights. While animal rights litigation has not had much success in the United States (Vayr, 2017, p. 849), it has obtained a few victories in Argentina, Colombia, and India (Peters, 2018, p. 356). These worldwide movements cast doubt on the idea that humans are the only class of legal subjects worthy of rights.

These trends speak to two existential crises facing humanity. First, the rise of robots in society calls into question the place of humans in the workforce and what it means to be human. By 2016, there were approximately 1.7 million robots working in industrial capacities and over 27 million robots deployed in professional and personal service roles, translating to around one robot per 250 people on the planet (van Oers & Wesselman, 2016, p. 5). The presence of robots is only likely to increase in the future, especially in service industries where physical work is structured and repetitive (Lambert & Cone, 2019, p. 6). Half of all jobs in the global economy are susceptible to automation, many of which may involve the use of robots designed to augment or replace human effort (Manyika et al., 2017, p. 5). In Japan, a labor shortage is driving businesses to utilize robots in occupations once the sole domain of humans, especially where jobs entail physically demanding tasks (Suzuki, 2019). The country's aging population is also accelerating the demand for robot assistance in elderly care (Foster, 2018). Some have questioned whether robots will come to replace humans in numerous fields such as, *inter alia*, agriculture (Jordan, 2018), journalism (Tures, 2019), manufacturing (Manyika et al., 2017), and medicine (Kocher & Emanuel, 2019). Others have argued that robots have and will continue to complement, not supplant, humans (Diamond, Jr., 2018).

The forward march to automate tasks currently assigned to humans for reasons related to economic efficiency, personal safety, corporate liability, and societal need is proceeding apace, while the ramifications of this shift are only beginning to be explored. One recent article suggests that the results of the 2016 U.S. presidential election may have been influenced to a non-trivial extent by the presence of industrial robots in certain labor markets (Frey et al., 2018). On a more philosophical level, advancements in technology, especially in the areas of artificial intelligence (AI) and robotics, have elicited discussions about the fundamental characteristics that define humans and the extent to which it might be possible to replicate them in synthetic form. What is it that makes humans special? Our intelligence? Memory? Consciousness? Capacity for empathy? Culture? If these allegedly unique characteristics can be reproduced in machines using complex algorithms, and if technology proceeds to the point where nonhuman entities are indistinguishable from their human counterparts, will this lead to the kind of destabilizing paradigm shift that occurred when Galileo confirmed the heliocentric theory of the universe?

Second, climate change threatens the existence of entire communities and invites reflection about the relationship between humans and nature. Despite the hope inspired by the widespread adoption of the Paris Climate Accord, recent estimates of the impact of Nationally Determined Contributions (NDCs) to the international agreement show that the world is on track to experience warming in excess of 3°C by 2100 (Climate Analytics, Ecofys and NewClimate Institute, 2018), a number well above the global goal of containing the rise in temperature to only 1.5°C. At the current rate of increasing temperatures, the planet is likely to reach the 1.5°C threshold between 2030 and 2052, with attendant impacts including sea-level rise, biodiversity loss, ocean acidification, and climate-related risks to agricultural or coastal livelihoods, food security, human health, and the water supply (IPCC, 2018). As such, climate change presents a clear and present danger not only to physical assets like lands and homes, but also to social institutions such as histories and cultures (Davies et al., 2017).

Acknowledgment of a changing climate and the degree to which it has been exacerbated by human activities has given rise to the idea that the Earth has transitioned from the Holocene to a new geological epoch—the Anthropocene (Crutzen, 2002; Zalasiewicz et al., 2007). Although some have taken issue with this proposal on the grounds that it masks the underlying causes responsible for the environmental changes observed (Haraway, 2015; Demos, 2017), others have found the concept useful for exploring the limitations of current systems and probing the boundaries of nature itself (Dodsworth, 2018). On the former point, Kotzé and Kim (2019) argue that the Anthropocene

> allows for an opening up of hitherto prohibitive epistemic "closures" in the law, of legal discourse more generally, and of the world order that the law operatively seeks to maintain, to a range of other understandings of, and cognitive frameworks for, global environmental change.
>
> (p. 3)

In this sense, the pronouncement of a new geological era offers an opportunity for critical examination of the law and how it might be reconceived to address the complex problems caused by industrialization. On the latter point, the Anthropocene renders human encounters with the natural world uncertain (Purdy, 2015, p. 230). It suggests the "hybridization of nature, as it becomes less and less autonomous with respect to human actions and social processes. To sustain a clear separation between these two realms is now more difficult than ever" (Arias-Maldonado, 2019, p. 51). More specifically, the Anthropocene presents a serious challenge to Cartesian dualism by rejecting ontological divisions in favor of a single, Latourian "flat" ontology defined by ongoing material processes, not static states of being (Arias-Maldonado, 2019, p. 53). In this reading of modernity, humans are both part of nature and act upon it (Dodsworth, 2018, p. 36). As a result, the boundary between humans and nonhumans has effectively collapsed.

The two trends—the development of machines made to look and act increasingly like humans, and the movement to recognize the legal rights of nonhuman "natural" entities—along with the two existential crises—the increasing presence of robots in work and social arenas, and the consequences of climate change and acknowledgment of humanity's role in altering the "natural" environment—lead us to revisit the question that is the focus of this book: *under what conditions might robots be eligible for rights?* Of course, a more appropriately tailored formulation might be—under what conditions might *some* robots be eligible for *moral or legal* rights? These italicized qualifications will prove important to the discussion in Chapter Two regarding the relationship between personhood and rights, and the interdisciplinary framework I put forth in Chapter Five that seeks to respond to the central question motivating this study. But before arriving at these key destinations, we need to first develop a common understanding about the kind(s) of technology relevant to the philosophical and legal analysis undertaken here.

Defining key terms

The word *robot* first entered the popular lexicon in Karel Čapek's 1921 play *R.U.R. (Rossum's Universal Robots)* (Čapek, 2004). Čapek based the term on the Czech word *robota*, which means "obligatory work" (Hornyak, 2006, p. 33). Interestingly, Rossum's robots were not machines at all, but rather synthetic humans (Moran, 2007). Today, however, robots have become almost universally associated with nonhuman machines. The International Organization for Standardization (ISO), for example, defines a "robot" as an "actuated mechanism programmable in two or more axes ... with a degree of autonomy ..., moving within its environment, to perform intended tasks" that is further classified as either industrial or service "according to its intended application" (International Organization for Standardization, 2012).

But this technical definition arguably fails to fully encapsulate the range of entities recognized as robots.[2] The "degree of autonomy" is perhaps ironic given the original definition's emphasis on servitude, and the performance of "intended tasks" seems to place a direct limit on the ability of a machine to act according to its own volition. Further, the ISO definition lacks any consideration of a robot's particular physical appearance or form. Winfield (2012) offers a more multifaceted definition that identifies robots according to their capabilities *and* form:

A robot is:

1. an artificial device that can sense its environment and purposefully act on or in that environment;
2. an embodied artificial intelligence; or
3. a machine that can autonomously carry out useful work. (p. 8)

The two elements coursing through this definition—capabilities and form—map nicely onto the debate over the machine question. Here we have three different capabilities—sensing, acting, and working autonomously—and three different forms—a device, an embodied AI, and a machine. As such, Winfield's conceptualization covers everything from a companion robot for the elderly to a mobile phone running an AI-based assistant to an industrial arm at a manufacturing facility. Later in his book, he fleshes out what he refers to as a "loose taxonomy" based on "generally accepted terms for classifying robots" (Winfield, 2012, p. 37). This classification system proposes six categories—mobility (fixed or mobile), how operated (tele-operated or autonomous), shape (anthropomorph, zoomorph, or mechanoid), human–robot interactivity, learning (fixed or adaptive), and application (industrial or service). As we shall see, several of these categories prove useful in distinguishing the types of robots that might warrant moral consideration.

But before proceeding, two other important terms must be adequately defined. First, what is an android, and how does it differ from a robot? The answer depends on the person responding to the question. For some in the science fiction community, *android* refers to "an artificial human of organic substance" (Stableford & Clute, 2019). This conceptualization resonates with Rossum's notion of robots, who were essentially humans grown in vats, but it could also apply to other popular examples such as Frankenstein's monster, or beings constructed out of the remains of past humans. For others, such as notable roboticist Hiroshi Ishiguro, androids are simply "very humanlike robot[s]" (Ishiguro, 2006, p. 320). Perhaps one of the more famous androids under this interpretation of the term is the character Data from the futuristic science–fiction television series *Star Trek: The Next Generation*. Thus, the definition of *android* seems to primarily revolve around the kind of materials constituting an entity, not its outward appearance. For the purposes of this book, *android* will refer to a synthetically produced human consisting of organic material, whereas *humanoid* will refer to a robot made of mechanical parts that is human-like in appearance (i.e., anthropomorphic in shape).

Second, what is AI? To be clear, as in the cases of *robot* and *android*, there is no consensus regarding the exact definition of AI. One group of definitions focuses on AI as a field of study. For instance, one author writes that AI is "a theoretical psychology … that seeks to discover the nature of the versatility and power of the human mind by constructing computer models of intellectual performance in a widening variety of cognitive domains" (Wagman, 1999, p. xiii). A panel of experts similarly conceives of AI as "a branch of computer science that studies the properties of intelligence by synthesizing intelligence" (Stone et al., 2016, p. 13). In bluntly practical terms, another scholar refers to AI as "the science of getting machines to perform jobs that normally require intelligence and judgment" (Lycan, 2008, p. 342). As an area of academic inquiry, AI comprises six disciplines—natural language processing, knowledge representation, automated reasoning, machine learning, computer vision, and robotics (Russell & Norvig, 2010, pp. 2–3). Importantly, robotics is seen as a discipline falling under the umbrella of AI, which suggests that intelligence is a necessary condition for objects to be considered robots.

A second (but related) group of AI definitions concerns the standards by which machines are adjudged to successfully approximate certain processes or behaviors. This group is further subdivided into definitions focused on the kind of process or behavior under scrutiny (i.e., thinking or acting) and the source of the standard being applied (i.e., human or rational) (Russell & Norvig, 2010, p. 1). Central to all of these definitions is the use of some kind of intelligence to accomplish certain tasks and an artefact (i.e., computer) that serves as the physical vehicle for the expression of intelligence. Notably, intelligence need not be determined by the extent to which an entity sufficiently emulates human reasoning; it can be compared against a measure of ideal performance. Although, like AI, *intelligence* has many definitions, one version of the concept that speaks to its application in computer science is "the ability to make appropriate generalizations in a timely fashion based on limited data. The broader the domain of application, the quicker conclusions are drawn with minimal information, the more intelligent the behavior" (Kaplan, 2016, pp. 5–6).

Generally speaking, experts distinguish between two types of AI—weak and strong. These types vary according to the degree to which artificial forms of intelligence prove capable of accomplishing complex tasks and the computer's ontological status based on the authenticity of its performance. In weak AI, the computer is "a system [designed] to achieve a certain stipulated goal or set of goals, in a manner or using techniques which qualify as intelligent" (Turner, 2019, p. 6). In strong AI, "computers given the right programs can be literally said to understand and have other cognitive states" (Searle, 1990, p. 67). In the former approach, the computer is merely a tool that generates the external appearance of intelligence; in the latter, the computer is an actual mind possessing its own internal states.

The weak versus strong AI debate hinges on whether computers simulate or duplicate mental states like those experienced by humans. Under a functionalist theory, engaging in processes like the manipulation of formal symbols is equivalent to thinking. In this account, mental states can be duplicated by a computer. Under a biological naturalist theory, on the other hand, there is something causally significant about processing information in an organic structure like the brain that makes thinking more than a sequence of translational tasks. Using this line of reasoning, at best, computers can only simulate mental states (Russell & Norvig, 2010, p. 954).

While René Descartes is credited with having been the first to consider whether machines could think (Solum, 1992, p. 1234), perhaps the most well-known illustrations of the extent to which computers might be able to demonstrate authentic intelligence were proposed by Alan Turing and John Searle. In Turing's (1950) imitation game, a human interrogator attempts to decipher the sex of two other players (one man and one woman), who are located in a separate room, by asking them a series of probing questions. Responses are then written and passed from one room to the other or communicated by an intermediary so as to avoid inadvertently compromising the game. The goal of the other players is to cause the interrogator to incorrectly guess their sex by offering clever responses. Turing then enquires about what would happen if a machine took the place of the man. He concludes that if a machine was able to successfully deceive the interrogator as often as a real human could, this would demonstrate that machines are effectively capable of thinking. This thought experiment thus suggests that behavior realistic enough to be indistinguishable from that exhibited by an organic person is functionally equivalent to the kind of thinking that we normally associate with humans.

As a rejoinder to Turing's test, Searle (1980) presented the "Chinese Room" argument (McGrath, 2011, p. 134). In this thought experiment, Searle imagines himself locked in a room where he receives a large amount of Chinese writing. Searle admittedly does not know any Chinese. He then receives a second delivery of Chinese writing, only this time it includes instructions in English (his mother tongue) for matching the characters in this batch with characters from the first batch. Finally, Searle obtains a third document written in Chinese that includes English language instructions on how to use the present batch to interpret and respond to characters in the previous two. After these exchanges, Searle also receives stories and accompanying questions in English, which he answers all too easily. Through multiple iterations involving the interpretation of Chinese characters, along with receipt of continuously improved instructions written by people outside the room, Searle's responses are considered indistinguishable from those of someone fluent in Chinese and just as good as his answers to the questions in English.

The important difference between the two tasks, according to Searle, is that he fully understands the English questions to begin with, while his responses to the Chinese questions are merely the product of mechanical symbol interpretation. This argument, *contra* Turing's, suggests that thinking requires more than executing tasks with high fidelity to a well-written program. Instead, thinking involves "intentionality," which is "that feature of certain mental states by which they are directed at or about objects and states of affairs in the world" (Searle, 2008, p. 333). It's not enough that inputs lead to the appropriate outputs; in order to qualify as being capable of thinking, a machine would need to possess mental states of its own that can be directed externally. Interestingly, Searle considers humans, by virtue of their capacity for intentionality, to be precisely the kind of machines one might accurately characterize as intelligent.

The present study is less concerned with resolving controversies regarding the definition of first-order concepts pertinent to AI and more interested in understanding how AI figures into the debate over which entities are deemed worthy of moral or legal consideration and, possibly, rights. Therefore, this book privileges definitions of AI that apply some standard of intelligence (be it human or ideal) to the processes or behaviors of technological artefacts. Although this approach might appear to sidestep the task of tethering the argument to a single, identifiable definition of AI, the reasons for doing so will become clear in the course of articulating a framework capable of assessing an entity's eligibility for rights. However, given that robotics is a discipline within the academic enterprise of AI, and provided that differences among robot types might affect the extent to which moral or legal status can be ascribed, presenting a definition of *robot* seems wise, if not essential. Therefore, for the purposes of this book, the term *robot* will apply to those nonhuman mechanical entities that operate under some form of AI and vary in capabilities and form according to mobility, how operated, shape, human–robot interactivity, learning, and application.[3]

A note on methodology

As a decidedly interdisciplinary endeavor, the quest to assess the eligibility of robots for certain rights beckons a methodological approach capable of providing insights from the page to the sage to the stage. In this spirit, the present text utilizes a range of qualitative methods, including philosophical analysis, comparative case studies, and structured interviews with robotics experts. During field work conducted in Tokyo, Japan, over the span of two weeks in August 2019, I completed six interviews with individuals in academia and the private sector. Two additional interviewees submitted responses to the questionnaire via email. Given the limited duration of my stay in Japan and the low number of study participants, I remain circumspect about drawing any definitive conclusions from these efforts. However, I do supplement the philosophical and legal analyses contained herein with occasional insights obtained through these interviews when appropriate. By attempting even a modestly multi-method project such as this one, I hope to illustrate the usefulness of applying "triangulation" (Jick, 1979, p. 602) to the study of roboethics, and encourage others to follow suit.

Contributions

This book makes three contributions to the study of rights in an era of great technological and environmental change. First, I offer fresh analyses intended to inform an answer to the machine question by drawing upon lessons from animal and environmental law. To date, a few scholars have written briefly about how developments in the RoN movement might influence the debate over robot rights (i.e., Gunkel, 2012; Torrance, 2013; Bryson et al., 2017; Turner, 2019). Some have also noted similarities between the machine question and the question as to whether or not animals should have rights (i.e., Coeckelbergh, 2011; Gunkel, 2012; Marx & Tiefensee, 2015; Hogan, 2017). However, none of the works listed above provides an extended analysis that examines both theory *and* practice regarding animal rights *and* the RoN. In addition, literature on the RoN has been curiously silent on the status of artefactual entities like robots. This book seeks to fill these gaps in the literature by bringing them into constructive dialogue with one another. Second, I present a new, multi-spectral framework for evaluating the conditions under which nonhuman entities might qualify for different forms of personhood, a precursor to rights. Bringing together heretofore disparate concepts and empirical evidence from anthropology, law, ethics, philosophy, and robotics, this tool offers academics, activists, judges, lawyers, and policymakers a context-dependent menu for assessing the extent to which intelligent machines might possess personhood(s). Third, I describe core tenets of a critical environmental ethic open to moral and legal recognition of nonhuman entities. This ethic derives inspiration from contemporary paradigm shifts observed across several disciplines, including the Anthropocene turn in philosophy (Arias-Maldonado, 2019), the materialist turn in the humanities and social sciences (Choat, 2017), the ontological turn in environmental law (Vermeylen, 2017), the relational turn in ethics (Coeckelbergh, 2010), and the relational turn in robotics (Jones, 2013). In so doing, I flesh out the practical implications of shifting to a "kincentric" (Salmón, 2000) and "posthuman" (Arvidsson, 2020, p. 123) ecological orientation.

Layout of the book

The remainder of the book proceeds as follows. In Chapter One, I review the literature on the moral status of robots. Scholars writing on this subject mainly fall into two camps—one focusing on the properties of an entity and the extent to which such properties qualify an entity as morally significant, and another emphasizing an entity's relations with humans or the larger socio-ecological context in which it operates. I close by elucidating the oversights and shortcomings of this debate, which include an inattention to the relationship between key terms and concepts (i.e., moral rights and legal rights), a blindness to (pro-Western) cultural biases that shape some of the arguments, and the inherent difficulty of addressing robot rights from within a single disciplinary silo (i.e., philosophical or legal).

In Chapter Two, I clarify the relationship between the muddled array of concepts central to the machine question that serve to justify or invalidate the basis for the possession of rights. In particular, I explain how aspects of the properties and relations approaches map onto different personhoods, statuses, and incidents. I also explore how theories of rights relate to the aforementioned approaches in an effort to distinguish alternate pathways to justifying the extension of rights. The purpose here is to disentangle the web of cross-listed terms and explicate defensible connections between them in order to provide a clear conceptual scheme that undergirds the framework presented later.

In Chapter Three, I examine the ways in which philosophical literature and case law on animal rights inform the discussion of rights for robots. I chronicle and analyze discussions about animal rights appearing in religious doctrine, Enlightenment thinking, philosophical treatises on animal ethics, and innovative legal theory. I also review the trials and tribulations of legal efforts to protect animal rights in courts across the United States, India, Argentina, and Colombia. From the foregoing evidence, I argue that relational approaches to animal rights present the strongest basis for affording animals enhanced protection, and that the success of animal rights appeals is highly context-dependent at present. I close by enumerating the conditions under which the animal rights model might advance the development of robot rights, including empirical verification of the presence of ontological properties, societal need for nonhuman personhood, and openness to non-Western ideas.

In Chapter Four, I detail how scholarship from environmental ethics, law, and philosophy, along with recent cases pertaining to the RoN, might provide a basis for extending rights to nonhuman entities. While environmental ethicists propose that the environment should be interpreted broadly to include all forms of life, analysts writing on critical environmental law, law in the Anthropocene, and New Materialism seek to disrupt conventional ideas about nature and agency, suggesting bolder imaginaries. I argue that the collapse of the human/nonhuman binary opens up the possibility of expanding the scope of rights. Next, I demonstrate how rights have already been extended to *natural* nonhuman entities under the auspices of the RoN, which have been adjudicated successfully in courts within Colombia, Ecuador, and India. Finally, from the foregoing evidence, I extract elements of a critical, Anthropocene-informed environmental law that support further widening of the concept of rights to include *artefactual* nonhuman entities, such as robots.

In Chapter Five, I stitch together insights from the preceding chapters to argue that rights can indeed be extended to robots. First, I return to the concepts defined and mapped out in Chapter Two. After considering the individual merits of specific properties or mechanisms, I demonstrate how both moral and legal personhoods/statuses/rights emerge from interactions between the two. As such, I contend that any solution to the machine question must necessarily take into account a combination of factors. Second, I probe the extent to which lessons from literature and litigation on animal rights translate to the context of robots. I find that the animal rights model illuminates important questions pertinent to the extension of rights to any nonhumans even if it does not supply the answers. Third, I argue that a critical, materialist, and broadly ecological interpretation of the environment, along with decisions by jurists establishing or upholding the RoN, support extension of such rights to nonhuman entities like robots. Fourth, I present a multi-spectral framework that can be used to assess whether or not different forms of technology (i.e., AI, algorithms, drones, robots, etc.) might be entitled to different types of personhood. Fifth, I suggest a praxis-oriented, critically inspired ethic that offers protective cover to intelligent machines. Sixth, I demonstrate the applicability of the aforementioned framework and ethic to hypothetical scenarios involving zoomorphic robot companions and anthropomorphic sex robots. Seventh, I close by suggesting areas for further research.

Notes

1 Scene from the film *Her* (Written and Directed by Spike Jonze, 2013, printed with permission from Annapurna Pictures).
2 As Gunkel (2018) explains, the term *robot* does not denote "some rigorously defined, singular kind of thing that exists in a vacuum. What is called 'robot' is something that is socially negotiated ... Its context (or contexts, because they are always plural and multifaceted) is as important as its technical components and characterizations" (p. 23).
3 Throughout this book, I use the terms *intelligent machines, intelligent artefacts, artificial agents, AI,* and *robots* interchangeably, notwithstanding certain qualifiers used in reference to specific types of robots (i.e., *humanoid robots*).

References

Arias-Maldonado, M. (2019). The "Anthropocene" in Philosophy: The Neo-material Turn and the Question of Nature. In F. Biermann & E. Lövbrand (Eds.), *Anthropocene Encounters: New Directions in Green Political Thinking* (pp. 50–66). Cambridge University Press.

Arvidsson, M. (2020). The Swarm That We Already Are: Artificially Intelligent (AI) Swarming 'Insect Drones', Targeting and International Humanitarian Law in a Posthuman Ecology. *Journal of Human Rights and the Environment, 11*(1), 114–137.

British Council (n.d.). *Should Robots Be Citizens?* British Council. Retrieved August 22, 2019, from https://www.britishcouncil.org/anyone-anywhere/explore/digital-identities/robots-citizens.

Bryson, J. J., Diamantis, M. E., & Grant, T. D. (2017). Of, for, and by the People: The Legal Lacuna of Synthetic Persons. *Artificial Intelligence and Law, 25*(3), 273–291.

Cano-Pecharroman, L. (2018). Rights of Nature: Rivers That Can Stand in Court. *Resources, 7*(1), 1–14.

Čapek, K. (2004). *R.U.R. (Rossum's Universal Robots)* (C. Novack, Trans.). Penguin.

Choat, S. (2017). Science, Agency and Ontology: A Historical-Materialist Response to New Materialism. *Political Studies, 66*(4), 1027–1042.

Climate Analytics, Ecofys and NewClimate Institute (2018, December 11). *The CAT Thermometer*. Climate Action Tracker. Retrieved from https://climateactiontracker.org/global/cat-thermometer/.

Coeckelbergh, M. (2010). Robot Rights? Towards a Social-Relational Justification of Moral Consideration. *Ethics and Information Technology, 12*(3), 209–221.

Coeckelbergh, M. (2011). Humans, Animals, and Robots: A Phenomenological Approach to Human-Robot Relations. *International Journal of Social Robotics, 3*(2), 197–204.

Crutzen, P. J. (2002). Geology of Mankind. *Nature, 415*(6867), 23.

Davies, K., Adelman, S., Grear, A., Magallanes, C. I., Kerns, T., & Rajan, S. R. (2017). The Declaration on Human Rights and Climate Change: A New Legal Tool for Global Policy Change. *Journal of Human Rights and the Environment, 8*(2), 217–253.

Demos, T. J. (2017). *Against the Anthropocene: Visual Culture and Environment Today*. Sternberg Press.

Diamond, Jr., A. M. (2018). Robots and Computers Enhance Us More Than They Replace Us. *The American Economist*, Online First. Retrieved from https://journals.sagepub.com/doi/abs/10.1177/0569434518792674.

Dodsworth, A. (2018). Defining the Natural in the Anthropocene: What Does the Right to a "Natural" Environment Mean Now? In M. Oksanen, A. Dodsworth, & S. O'Doherty (Eds.), *Environmental Human Rights: A Political Theory Perspective* (pp. 33–46). Routledge.

Foster, M. (2018, March 27). Aging Japan: Robots May Have Role in Future of Elder Care. *Reuters*. Retrieved from https://www.reuters.com/article/us-japan-ageing-robots-wider image-idUSKBN1H33AB.

Frey, C. B., Berger, T., & Chen, C. (2018). Political Machinery: Did Robots Swing the 2016 US Presidential Election? *Oxford Review of Economic Policy, 34*(3), 418–442.

Global Alliance for the Rights of Nature (2019). *Explore the Rights of Nature Around the World*. Global Alliance for the Rights of Nature. Retrieved from https://therightsofnature.org.

Gunkel, D. J. (2012). *The Machine Question: Critical Perspectives on AI, Robots, and Ethics*. MIT Press.

Gunkel, D. J. (2018). *Robot Rights*. MIT Press.

Haraway, D. (2015). Anthropocene, Capitalocene, Plantationocene, Chthulucene: Making Kin. *Environmental Humanities*, *6*(1), 159–165.

Hart, R. D. (2018, February 14). Saudi Arabia's Robot Citizen is Eroding Human Rights. *Quartz*. Retrieved from https://qz.com/1205017/saudi-arabias-robot-citizen-is-eroding-human-rights/.

Hatmaker, T. (2017, October 26). Saudi Arabia Bestows Citizenship on a Robot Named Sophia. *Techcrunch*. Retrieved from http://social.techcrunch.com/2017/10/26/saudi-arabia-robot-citizen-sophia/.

Hi, I am Sophia... (2019). Hanson Robotics. Retrieved from https://www.hansonrobotics.com/sophia/.

Hogan, K. (2017). Is the Machine Question the Same Question as the Animal Question? *Ethics and Information Technology*, *19*(1), 29–38.

Hornyak, T. N. (2006). *Loving the Machine: The Art and Science of Japanese Robots*. Kodansha International.

IPCC (2018). Summary for Policymakers. In V. Masson-Delmotte, P. Zhai, H.-O. Pörtner, D. Roberts, J. Skea, P. R. Shukla, A. Pirani, W. Moufouma-Okia, C. Péan, R. Pidcock, S. Connors, J. B. R. Matthews, Y. Chen, X. Zhou, M. I. Gomis, E. Lonnoy, T. Maycock, M. Tignor, & T. Waterfield (Eds.), *Global Warming of 1.5°C: An IPCC Special Report on the Impacts of Global Warming of 1.5°C Above Pre-industrial Levels and Related Global Greenhouse Gas Emission Pathways, in the Context of Strengthening the Global Response to the Threat of Climate Change, Sustainable Development, and Efforts to Eradicate Poverty*. Retrieved from https://www.ipcc.ch/sr15/chapter/spm/

Ishiguro, H. (2006). Android Science: Conscious and Subconscious Recognition. *Connection Science*, *18*(4), 319–332.

ISO (2012). *ISO 8373:2012(en), Robots and Robotic Devices—Vocabulary*. Online Browsing Platform. Retrieved from https://www.iso.org/obp/ui/#iso:std:iso:8373:ed-2:v1:en.

Jick, T. D. (1979). Mixing Qualitative and Quantitative Methods: Triangulation in Action. *Administrative Science Quarterly*, *24*(4), 602–611.

Jones, R. A. (2013). Relationalism through Social Robotics. *Journal for the Theory of Social Behaviour*, *43*(4), 405–424.

Jonze, S. (2013). *Her*. Annapurna Pictures.

Jordan, M. (2018, November 20). As Immigrant Farmworkers Become More Scarce, Robots Replace Humans. *New York Times*. Retrieved from https://www.nytimes.com/2018/11/20/us/farmworkers-immigrant-labor-robots.html.

Kaplan, J. (2016). *Artificial Intelligence: What Everyone Needs to Know*. Oxford University Press.

Kauffman, C. M., & Martin, P. L. (2018). Constructing Rights of Nature Norms in the US, Ecuador, and New Zealand. *Global Environmental Politics*, *18*(4), 43–62.

Kocher, B., & Emanuel, Z. (2019, March 5). Will Robots Replace Doctors? *Brookings Institution*. Retrieved from https://www.brookings.edu/blog/usc-brookings-schaeffer-on-health-policy/2019/03/05/will-robots-replace-doctors/.

Kotzé, L. J., & Kim, R. E. (2019). Earth System Law: The Juridical Dimensions of Earth System Governance. *Earth System Governance*, *1*, 1–12.

Lambert, J., & Cone, E. (2019). *How Robots Change the World: What Automation Really Means for Jobs and Productivity*. Oxford Economics. Retrieved from http://resources.oxfordeconomics.com/how-robots-change-the-world.

Lycan, W. G. (2008). Robots and Minds. In J. Feinberg & R. Shafer-Landau (Eds.), *Reason and Responsibility: Readings in Some Basic Problems of Philosophy* (13th ed., pp. 342–348). Thomson Wadsworth.

Manyika, J., Chui, M., Miremadi, M., Bughin, J., George, K., Willmott, P., & Dewhurst, M. (2017). *A Future That Works: Automation, Employment, and Productivity*. McKinsey Global Institute. Retrieved from https://www.mckinsey.com/~/media/McKinsey/Fea tured%20Insights/Digital%20Disruption/Harnessing%20automation%20for%20a %20future%20that%20works/MGI-A-future-that-works_Executive-summary.ashx.

Marx, J., & Tiefensee, C. (2015). Of Animals, Robots and Men. *Historical Social Research / Historische Sozialforschung, 40*(4), 70–91.

McGrath, J. F. (2011). Robots, Rights, and Religion. In J. F. McGrath (Ed.), *Religion and Science Fiction* (pp. 118–153). Pickwick.

Moran, M. E. (2007). Rossum's Universal Robots: Not the Machines. *Journal of Endourology, 21*(12), 1399–1402.

Peters, A. (2018). Rights of Human and Nonhuman Animals: Complementing the Universal Declaration of Human Rights. *AJIL Unbound, 112*, 355–360.

Purdy, J. (2015). *After Nature: A Politics for the Anthropocene*. Harvard University Press.

Russell, S. J., & Norvig, P. (2010). *Artificial Intelligence: A Modern Approach* (3rd ed.). Pearson Education.

Salmón, E. (2000). Kincentric Ecology: Indigenous Perceptions of the Human-Nature Relationship. *Ecological Applications, 10*(5), 1327–1332.

Searle, J. R. (1980). Minds, Brains, and Programs. *Behavioral and Brain Sciences, 3*(3), 417–457.

Searle, J. R. (1990). Minds, Brains, and Programs. In M. A. Boden (Ed.), *The Philosophy of Artificial Intelligence* (pp. 67–88). Oxford University Press.

Searle, J. R. (2008). Minds, Brains, and Programs. In: J. Feinberg & R. Shafer-Landau (Eds.), *Reason and Responsibility: Readings in Some Basic Problems of Philosophy* (13th ed., pp. 330–342). Thomson Wadsworth.

Solum, L. B. (1992). Legal Personhood for Artificial Intelligences. *North Carolina Law Review, 70*(4), 1231–1288.

Stableford, B. M., & Clute, J. (2019). *Androids* (J. Clute, D. Langford, P. Nicholls, & G. Sleight Eds.). Gollancz. Retrieved from http://www.sf-encyclopedia.com/entry/an droids.

Stone, P., Brooks, R., Brynjolfsson, E., Calo, R., Etzioni, O., Hager, G., Hirschberg, J., Kalyanakrishnan, S., Kamar, E., Kraus, S., Leyton-Brown, K., Parkes, D., Press, W., Saxenian, A., Shah, J., Tambe, M., & Teller, A. (2016). Artificial Intelligence and Life in 2030. In *One Hundred Year Study on Artificial Intelligence: Report of the 2015– 2016 Study Panel*. Stanford University. Retrieved from http://ai100.stanford.edu/2016 -report.

Suzuki, W. (2019, February 8). At Your Service: Japanese Robots Move out of the Factory. *Nikkei Asian Review*. Retrieved from https://asia.nikkei.com/Business/Business-trends/ At-your-service-Japanese-robots-move-out-of-the-factory.

Torrance, S. (2013). Artificial Agents and the Expanding Ethical Circle. *AI and Society, 28*(4), 399–414.

Tures, J. A. (2019, July 14). J Robot: Could Artificial Intelligence Actually Replace Reporters? *Observer*. Retrieved from https://observer.com/2019/07/journalism-robots -reporters-jobs-artificial-intelligence/.

Turing, A. M. (1950). I.—Computing Machinery and Intelligence. *Mind, 59*(236), 433–461.

Turner, J. (2019). *Robot Rules: Regulating Artificial Intelligence*. Palgrave Macmillan.

van Oers, R., & Wesselman, E. (2016). *Social Robots*. KPMG. Retrieved from https://as sets.kpmg/content/dam/kpmg/pdf/2016/06/social-robots.pdf.

Vayr, B. (2017). Of Chimps and Men: Animal Welfare vs. Animal Rights and How Losing the Legal Battle May Win the Political War for Endangered Species. *University of Illinois Law Review*, *2*, 817–876.

Vermeylen, S. (2017). Materiality and the Ontological Turn in the Anthropocene: Establishing a Dialogue Between Law, Anthropology and Eco-Philosophy. In L. J. Kotzé (Ed.), *Environmental Law and Governance for the Anthropocene* (pp. 137–162). Hart Publishing.

Vincent, J. (2017, October 30). Pretending to Give a Robot Citizenship Helps No One. *The Verge*. Retrieved from https://www.theverge.com/2017/10/30/16552006/robot-rights -citizenship-saudi-arabia-sophia.

Wagman, M. (1999). *The Human Mind According to Artificial Intelligence: Theory, Research, and Implications*. Praeger.

Winfield, A. (2012). *Robotics: A Very Short Introduction*. Oxford University Press.

Zalasiewicz, J., Williams, M., Smith, A., Barry, T. L., Coe, A. L., Brown, P. R., Brenchley, P., Cantrill, D., Gale, A., Gibbard, P., Gregory, F. J., Hounslow, M. W., Kerr, A. C., Pearson, P., Knox, R., Powell, J., Waters, C., Marshall, J., Oates, M., … Stone, P. (2007). Are We Now Living in the Anthropocene? *Geological Society of America Today*, *18*(2), 4–8.

1 Rights for robots

Making sense of the machine question

Sometimes I Forget You're a Robot

(Sam Brown, 2013)

Most of the literature on the ethical dimensions of robots concerns at least one of the five following areas: (1) human actions completed through the use of robots, (2) the moral standing of robots, (3) the behavior of robots, (4) the ethical implications of introducing robots into social or occupational spaces, and (5) self-reflection by scholars regarding the impact of robots on their field of study (Steinert, 2014, p. 250). In this book, I am primarily interested in contributing to the second area of inquiry listed above (along with its analog in the legal domain), although this is not to diminish the importance of any of the other ethical issues raised by robots and their application in human endeavors. For instance, there is exciting and important research being conducted on the ethics of drone warfare (i.e., Enemark, 2013), how robots deployed in nursing homes act towards the elderly (i.e., Sharkey & Sharkey, 2012), the effects of using robots in the classroom on teachers and children (i.e., Serholt et al., 2017), ethical considerations in the design of robots used for love or sex (i.e., Sullins, 2012), and the ethical conduct of scholars working on human–robot interaction (HRI) (i.e., Riek & Howard, 2014). The point here is that the discussion regarding the field of "roboethics" (Veruggio & Operto, 2006, p. 4) is far more complicated and multi-faceted than is suggested by the narrow slice entertained in this work. We have come a long way from Asimov's (1942) three laws of robotics, which exclusively prescribed ethical directives intended to govern robot behavior.

The present text focuses on the moral and legal standing of robots, and seeks to develop a response to the following question—*can robots have rights?* This line of inquiry necessarily entails five separate, albeit related, sub-questions:

(i) Which *kinds of robots* deserve rights? (ii) Which *kinds of rights* do these (qualifying) robots deserve? (iii) Which *criterion*, or cluster of criteria, would be essential for determining when a robot could qualify for rights? (iv) Does a robot need to satisfy the conditions for (moral) *agency* in order to qualify for at least some level of moral consideration? (v) Assuming that certain kinds of robots may qualify for some level of moral consideration, which *kind of rationale* would be considered adequate for defending that view?

(Tavani, 2018, p. 1; emphasis in original)

Throughout this work, each of these sub-questions will be answered to some extent. As advance warning, more effort will be expended to identify the kinds of robots that might deserve rights, establish the criterion for determining rights eligibility, assess the importance of agency in the calculation of moral consideration, and explain the rationale invoked to support the preceding arguments than to itemize specific rights that might be bestowed upon robots.

Framing the debate: Properties versus relations

Broadly speaking, ethicists, philosophers, and legal scholars have extensively debated the answer to the machine question, with some finding that robots might qualify for rights and others rejecting the possibility on jurisprudential, normative, or practical grounds. Both sides of the debate frame their positions chiefly in terms of either the properties of an intelligent machine or its relationship to other entities (Tavani, 2018, p. 2). This division has its roots in the philosophical concept known as the is/ought problem, articulated by Hume (1738/1980) in *A Treatise of Human Nature*. The problem, so to speak, occurs when a value-laden statement masquerades as a fact-based one; we treat something a certain way by virtue of how we think it *ought* to be treated, not by virtue of what it actually *is*. Therefore, the philosophical task of figuring out the moral status of an entity and how to act towards it necessarily involves understanding whether *ought* is derived from *is* or vice versa.[1] More concretely, in the properties-based approach, the way we decide how to treat a robot (how we believe we *ought* to engage with it) depends on its characteristics (what it *is*). In the relational approach, the moment we enter into social relations with an entity, obligations towards it are established (how we *ought* to treat it) irrespective of the qualities that suggest its alterity (what it *is*).[2] In the space here, I briefly summarize the thrust of these arguments with an eye towards more fully examining the relationship between these positions and cognate concepts such as personhood and rights, which I discuss in Chapter Two. As we shall see, the lively discussion about robot rights has suffered from an inattention to the relationship between key concepts, unacknowledged cultural biases, and challenges associated with tackling an interdisciplinary problem.

One camp consists of analysts who argue that robots do not or should not have rights, focusing mainly on the properties of such intelligent artifacts and, to a lesser extent, on the relational dimension of HRI. In one of the earlier works indicative of this perspective, Miller (2015) contends that what separates humans and animals from "automata" is the quality of "existential normative neutrality" (p. 378). Whereas the ontological status of humans and animals is taken for granted, the existence of automata is actively constructed by human agents. Confusingly, Miller writes about the connection between *moral* status and the eligibility for full *human* rights, by which he means the entire suite of *legal* rights expressed in major international human rights documents. In addition, he claims that "humans are under no moral obligation to grant full human rights to entities possessing ontological properties critically different from them in terms of human rights bases" (Miller, 2015, p. 387). This assertion nearly qualifies as a strawman argument. As shown below, those finding robot rights philosophically tenable do not advocate for the assignment of all major human rights to technological entities. Furthermore, conflating *moral* rights with *legal* rights overlooks the varied reasons why nonhumans might be and have been extended the latter kind of protection.

For Solaiman (2017), the question revolves around the extent to which robots can fulfill *legal* duties, which are "responsibilities commanded by law to do or to forbear something for the benefit of others, the failure in, or disobedience of, which will attract a remedy" (p. 159). Whereas corporations consist of people who can perform duties and idols have managers who tend to their legal interests, robots have no such human attachments. Therefore, since robots cannot fulfill *legal* duties, they cannot meet the criteria for *legal* personhood and thus they are not entitled to *legal* rights.

Bryson et al. (2017) rebuff the idea of granting either *moral* or *legal* rights to robots. They contend that robots do not possess the qualities intrinsic to moral patients (i.e., consciousness), so they cannot hold *moral* rights or be considered *moral* patients, making them ineligible for *legal* personhood, and thus not entitled to *legal* rights (pp. 283–4). Further, leaning on Solaiman, the authors urge that absent the ability to be held accountable for one's actions, an artificial entity cannot fulfill *legal* duties and therefore does not qualify as a *legal* person. This lack of accountability could result in "humans using robots to insulate themselves from liability and robots themselves unaccountably violating human legal rights" (Bryson et al., 2017, p. 285).[3] Neither of these outcomes advance the ultimate objective of an established legal order—"to protect the interests *of* the people" (Bryson et al., 2017, p. 274; emphasis in original). In short, the costs of affording robots rights outweigh the benefits of doing so.

For Bryson (2018), robots should not be assigned the status of either *moral* agents or *moral* patients because doing so would place human interests in competition with the interests of artificial entities, which is unethical. Determining whether an entity qualifies as a *moral* patient or a *moral* agent is critical in establishing whether or not it possesses *moral* duties and/or *moral* rights. Bryson agrees with Solaiman that while humans have the power to assign *legal* duties and *legal* rights to any entity, these forms of recognition are only available to "agent[s] capable of knowing those rights and carrying out those duties" (Bryson, 2018, p. 16). If a robot does not meet the criteria for either *moral* agency or *moral* patiency, it cannot hold *moral* rights.[4] In fact, Bryson (2010) contends controversially, robots should be treated as mere slaves.[5]

More recently, Birhane and van Dijk (2020) adopt a "post-Cartesian, phenomenological view" and conclude that "robots are [not] the kinds of beings that could be granted or denied rights" (p. 2). Whereas all humans share a capacity for "lived embodied experience" (Birhane & van Dijk, 2020, p. 2), robots do not. Robots are technological artefacts that may contribute to the human experience, but they are merely elements present in the human social world, not beings unto themselves. As such, the authors take a relational approach to robot rights but reach a conclusion totally opposite from the one obtained by Coeckelbergh (2010, 2011, 2014) and Gunkel (2012, 2018a).[6] Finally, instead of focusing on the rights of robots, the scholars suggest, we should concentrate our efforts on safeguarding human welfare, which is the ultimate reason for contemplating rights for AI anyway.

This article is logically flawed and deeply contradictory, rendering its arguments highly suspect. First, the very title of the piece frames the issue in terms of both a strawman argument and a false dichotomy. Robot rights are neither promoted solely as a means of advancing human welfare, nor are robot rights and human welfare mutually exclusive objectives. Second, their alleged employment of a post-Cartesian outlook is belied by their assessment that while robots are embedded in human social practices, they are still different enough from humans to warrant their exclusion from the moral circle. This move ignores the ontological flattening that occurs when viewing the moral universe as a social-relational whole. If, in fact, "technologies are always *already part of ourselves*" (Birhane & van Dijk, 2020, p. 3; emphasis in original), there is no basis for the kind of ontological separation described by Descartes. In short, the authors fail to present a convincing case for the dismissal of robot rights.

Another camp comprises those writers who maintain that robots could conceivably possess rights, exploring the possibilities generated by the properties of such entities, their relationship with humans and the larger context in which they operate, or a combination of the two. The justifications supplied by these advocates are mostly philosophical, but a few are explicitly legal in nature. For the moment, I leave aside arguments that do not directly engage with the question of rights (i.e., those dealing primarily with concepts like intentionality, personhood, and being alive).[7]

On the properties side, Chao (2010) claims that *legal* rights should be extended only to "fully rational" robots that exhibit "consciousness, intentionality, and free will" because to deny them such protections in light of their possession of such characteristics would be "inconsistent" with the standard by which humans are granted rights (p. 98). Hubbard (2011), also seeking to maintain the logic governing human attributions of elevated *moral* status, argues that machines capable of complex intellectual interaction, self-consciousness, and living in a community on the basis of reciprocal self-interests should be given "the basic Lockean right of self-ownership" (p. 417). McGrath (2011) suggests that the designation of rights depends on determining whether or not a machine is sentient (p. 139). Marx and Tiefensee (2015) answer the machine question in terms of how well intelligent artefacts approximate human qualities: "[i]n order to be regarded as the holder of rights, robots would have to be sentient beings with an idea of a subjective good and important interests that are worthy of protection" (p. 85). However, they also offer the caveat that if there is no meaningful moral difference between how humans and robots feel pain, robots might be afforded rights. Danaher (2020) advances a theory of ethical behaviorism—*moral* status should be based on observable behaviors and reactions towards humans. Because mental states are unobservable,[8] external behaviors constitute the only accessible means of establishing whether or not an entity possesses the kinds of metaphysical properties required to obtain *moral* status (i.e., consciousness or intelligence). Therefore, determinations of *moral* rights and duties can be achieved by perceiving behavior that is performatively equivalent to that attributed to other entities deemed worthy of moral consideration by virtue of their properties.

On the relational side, positions may be further classified as social or ecological in orientation (Jones, 2013). The social position emphasizes the interaction that a robot has with another entity or the mental representations produced through an encounter with an Other. The ecological position, on the other hand, considers the extent of an entity's embeddedness within a culture, how its embodiment structures perceptions and physical responses, or the degree to which it stands in harmony with all things. Levy (2009) approaches the subject from a social perspective, arguing that despite not being sentient like animals, intelligent machines still might deserve rights because how we treat them will mirror how humans treat one another. He concludes by instructing that "treating robots in ethically suspect ways will send the message that it is acceptable to treat humans in the same ethically suspect ways" (Levy, 2009, p. 215). Coeckelbergh (2010) advocates in favor of a "social ecology" that rejects *a priori* distinctions between human and nonhuman entities, and instead promotes moral consideration based on experiences and the contexts in which they occur (p. 217). Interestingly, Coeckelbergh does not foreclose the possibility that properties may play a role in a relational approach to moral consideration. Instead, he leaves room for "properties-as-they-appear-to-us within a social-relational, social-ecological context" (Coeckelbergh, 2010, p. 219). The same author later invokes both the social and ecological (i.e., embodiment) perspectives of a relational approach when he writes, "what matters for understanding and evaluating human–robot relations is how the robot appears to us" (Coeckelbergh, 2011, p. 198).[9] In another work, Coeckelbergh (2014) also acknowledges the importance of an entity's ecological embeddedness: "[o]ur personal construction of the robot is influenced by the way our culture constructs machines, and this construction is not only a word process but also a living process" (p. 69). Darling (2016) presents an exclusively social-relational justification for extending legal protections to social robots. She argues that the human tendencies to anthropomorphize nonhumans, project our emotions onto them, and shield ourselves from the emotional harm we experience when witnessing the abuse of other entities suggest that we ought to regulate violent behavior towards social robots through laws similar to animal welfare statutes. The litany of sympathetic user comments posted in response to a Boston Dynamics (2016) YouTube video in which a man strikes a humanoid robot with a hockey stick reflects precisely the kind of sensitivity towards nonhumans identified by Darling. She stops short, however, of advocating for robot rights, which she regards as fodder for philosophical musings, not policymaking (at least for now).

Although plenty of philosophers have proposed general theories of *moral* status that address both human and nonhuman entities (i.e., Warren, 1997; Fox, 2006;

Metz, 2012), few have endeavored specifically to prescribe the boundaries of the moral circle with robots in mind. Floridi (1999, 2008) has attempted such an all-encompassing ecological rendering of the ethical universe. In Floridi's (1999) concept of "Information Ethics" (IE), the line delineating those entities worthy of moral consideration from those that do not warrant such treatment pushes outwards from its initial encircling of strictly physical beings like animals, persons, and plants to ultimately include "every instance of information, no matter whether physically implemented or not" (p. 43). Under an IE vision, the individual identifying factors associated with an entity are conceived as data structures, while behaviors or reactions are governed by a grouping of functions, operations, or procedures. Anything that is a kind of information or simply *is* possesses intrinsic value. In more concise terms, "life" is replaced with "existence" (Floridi, 2008, p. 60). Any entity that suffers from entropy, here defined as "any kind of destruction or corruption of informational objects ... that is, any form of impoverishment of being, including nothingness" (Floridi, 2008, p. 60), is deserving of moral concern. By making this kind of shift from *bio*centrism to *onto*centrism, certain kinds of robots and even non-embodied AI clearly earn a place within the moral circle by virtue of their beingness and entropic condition. However, IE is vulnerable to the criticism that it is perhaps over-inclusive, placing excessive duties on entities capable of contributing to the destruction of any and all information-based beings.

Gunkel (2018a), drawing on the work of philosopher Emmanuel Levinas and largely concurring with Coeckelbergh, develops a social-relational approach he refers to as "thinking otherwise" (p. 159). Here, the *moral* status of an entity emerges from an encounter with it, obligating us to respond to its presence before we fully understand its inner workings or capacities. The determination regarding the kind of entity we face follows from this initial ethical reaction. In Humean terms, "*ought* precedes and takes precedence over *is*" (Gunkel, 2018a, p. 166; emphasis in original). Gunkel not only rejects the properties approach, but he also disfavors the strategy employed by information and environmental ethicists whereby the moral circle is progressively broadened in an effort to include more and more kinds of entities within its ambit. He argues that "these different ethical theories endeavor to identify what is essentially the same in a phenomenal diversity of different individuals. Consequently, they include others by effectively stripping away and reducing differences" (Gunkel, 2018a, p. 163). This interpretation of "competing centrisms" (Gunkel, 2018a, p. 163) is not quite accurate, however, insofar as it relates to some of the more ecologically minded ethical or legal approaches, which explicitly call for the ontological de-centering of humans (i.e., Philippopoulos-Mihalopoulos, 2011, 2017; Vermeylen, 2017; Tavani, 2018, pp. 401–402).

(Dis)integrative approaches

Some writers have attempted to unify the properties and relational approaches in service of expanding the moral circle to include information technology-based entities. Søraker (2007) propounds a relational theory that assesses *moral* status on the basis of both intrinsic and relational properties. Intrinsic properties refer to qualities like free will, reason, and self-consciousness, while two relational properties that work in tandem—irreplaceability and constitutivity—are also identified. Unpacking this argument, the author holds that "an entity that is an irreplaceable and constitutive part of an organic unity together with a person thereby attains value as an end and moral status" (Søraker, 2007, p. 16). While relational properties might be able to confer moral *status* on an entity, only through the demonstration of intrinsic properties can a being enjoy full moral *standing*. Thus, *moral* status may be a matter of degree, not an all-or-nothing proposition.[10] The resulting integrationist model ordains a hierarchy that arranges the following entities in descending order of moral significance: moral persons, merely self-conscious beings, merely sentient beings, and non-sentient entities. Although this effort intends to bring both properties and relations into the same moral calculus, it clearly privileges the former at the expense of the latter. Humans are still at the apex of the moral order, and the extent to which nonhuman entities might obtain *moral* status is still somewhat dependent on the instrumental value that such entities hold for *moral* persons.

A charitable reading of Coeckelbergh (2012) suggests that he, too, strives for an integrationist solution to the properties-versus-relations debate. Expanding on the notion of appearances present in his earlier work, Coeckelbergh offers the important insight that an entity may be viewed in multiple ways depending on the context and how humans relate to it. This phenomenological approach to determining *moral* status rejects ontological stability, recognizing that the reality of an object is inextricably linked to the subject which encounters it. The logical application of this "deep-relational" approach to the domain of AI and robotics entails that

> [a]n intelligent humanoid robot may appear as a machine (an object, a thing) but also as a living tool; or it may appear as a human, an other (a social other) or a subject. It may even appear as a companion, partner, friend, and so on.
> (Coeckelbergh, 2012, pp. 45, 44)

Although Coeckelbergh does not avoid the charge of anthropocentrism, he successfully retreats from a properties-centric approach and leaves room for multiple ontologies and direct experiences in the process of assigning *moral* status.

Others have rejected both the properties and relational approaches. Tavani (2018) takes issue with the properties approach on the grounds that it "fails to show us why one property or set of properties should be privileged over another" and finds fault in the relational approach, which "leaves us with many unresolved questions" (p. 2). The author, agreeing with Gunkel (2018b), contends that robots might warrant moral consideration as moral *patients*, but not moral *agents*. Relying on Jonas' (1984) update of Heidegger's concept of being-in-the-world, Tavani (2018) argues that humans have moral obligations to beings with a significant presence in the technological-world, such as social robots (p. 12). Curiously, while Tavani explicitly locates his position outside the realm of the relational approach, as indicated above, this perspective includes an ecological sensitivity that is wholly compatible with the idea of robots as beings-in-the-technological-world. As such, he does not manage to successfully evade identification with one of the two camps in the machine question debate.

Rounding out this discussion, Torrance (2013) presents a typology useful for categorizing the aforementioned properties or relational approaches according to the kind of object positioned at the center of the moral circle. Specifically, he identifies four macroethical perspectives—anthropocentrism, biocentrism, eco-centrism, and infocentrism. Anthropocentrism privileges humans, human-like qualities, human interests and needs, or relations with humans. Biocentrism holds all organic beings with biologically determined capabilities in higher esteem than any created through synthetic means. Ecocentrism prioritizes whole natural ecosystems or individual organic beings present in natural ecosystems.[11] Finally, infocentrism interprets characteristics associated with intelligence or the mind as integral to deciding the moral status of an entity.[12]

Gunkel (2018a, p. 163) decries these perspectives as essentially exercises in progressively larger circle drawing and situates his own argument outside of this competition, focusing instead on the spontaneous emergence of relations that occurs when encountering an Other of *any* material composition. However, both he and Coeckelbergh seem to implicitly deny the relevance of features that shape the interaction between object and subject. For instance, there is a substantial body of evidence demonstrating that humans unconsciously apply social expectations to and form social relationships with technological artifacts like computers and televisions (Reeves & Nass, 1996; Nass & Moon, 2000). The conditions under which forms of technology elicit social responses remains an unsettled issue, but it is evidently important to understanding the resulting interactive behavior observed in humans. More bluntly, it seems likely that the perception that an entity possesses lifelike qualities motivates humans to treat Apple's Siri differently from the way we treat less animated or dynamic things like a static image or a boulder.

No matter how much analysts want to champion a particular side in this debate, or eschew both properties and relations altogether, it stands to reason that any approach to assessing the *moral* status of robots will have to at least consider the characteristics of the entity in question along with the context in which it exists. In this book, I am not angling to resolve longstanding philosophical disputes or prescribe a superior macroethical orientation that specifies the boundaries of the moral circle for all entities present in this world. The intent of this project is more circumscribed—I wish to propose a method of assessing the eligibility of certain kinds of robots for personhood and thus rights. Any claim that the framework developed here might apply to other circumstances must be made by the reader (be they human or AI).

Shortcomings of the debate

As indicated at the outset of this chapter, the discourse regarding the machine question is marred by a few important shortcomings. First, scholars have paid insufficient attention to the relationship between key concepts. As evidenced by the use of italics above, some of the authors cited discuss *legal* duties/personhood/status/rights while others focus on *moral* duties/personhood/status/rights. Still others oscillate between legal and moral terms. Yet rarely do they explain how, if at all, these concepts are connected to one another. To be sure, "[m]oral rights are not the same as legal rights, though protection in law often follows shortly after society has recognised a moral case for protecting something" (Turner, 2019, p. 170). In the next chapter, I attempt to overcome this terminological morass by clearly articulating the links between properties or relational mechanisms, personhoods, statuses, and rights.

Second, some of the literature in this area lacks the kind of self-reflexiveness that encourages acknowledging one's own cultural biases. Hard and fast conclusions regarding what is or is not moral, what should or should not be legal, and what does or does not count as a person ignore the multiplicity of perspectives and ontologies found across space and time. For instance, there exists a discernible difference of opinion among Western and Eastern philosophical traditions regarding the social standing of artifacts (Mathews, 1991). As will be argued throughout this book, the West does not own a monopoly on the truth, and Eastern and Indigenous ways of thinking about nonhuman entities deserve a place in the conversation about robot rights.

Third, some of the inconsistencies and oversights present in this burgeoning area of inquiry are merely the product of the interdisciplinary nature of the subject matter. Asking the machine question invites many different kinds of experts to the table—cognitive scientists, engineers, ethicists, futurists, historians, journalists, legal scholars, philosophers, social scientists, and roboticists, among others. None of these groups on their own can legitimately lay claim to the entirety of knowledge on the subject of rights for robots. As such, any effort to push this discussion forward should explicitly indicate the limitations of writing from a given disciplinary vantage point and draw upon the work of those in cognate fields to the extent possible. As a political scientist and sometimes legal scholar, I fully recognize the depth of the pool I have elected to wade into. Any factual inaccuracies or technological misunderstandings contained within this book are solely my own.

To continue this journey, we must first attempt to resolve some of the terminological inconsistencies and conceptual disconnects that frustrate the ability of scholars from different fields to engage in meaningful dialogue about the audacious possibility of robot rights. Such is the task that lies ahead in Chapter Two.

Notes

1 That is, if debating moral status is indeed the appropriate manner in which to proceed. One philosopher argues that the very framing of ethical questions regarding marginal cases in terms of moral status is ultimately unproductive, as the discussion could be conducted using simpler terms. See Sachs (2011).
2 For an extensive treatment of the is/ought problem in the context of robot rights, see Gunkel (2018a).

3 But see Turner (2019), who argues that "AI personality allows liability to be achieved with minimal damage to fundamental concepts of causation and agency, thereby maintaining the coherence of the system as a whole" (p. 186).

4 The computer scientist and ethicist's objection to bestowing rights upon robots is unequivocal: "If robots ever need rights we'll have designed them unjustly" (Bryson, 2017). Of course, this strongly worded statement ignores the fact that robot design is determined by, *inter alia*, their functional problem-solving capabilities, the intellectual aspirations of roboticists, market potential, national research priorities, and realizing forms adapted from science fiction. Danaher (2020) suggests that the impulse to design robots capable of performing in ways similar to other entities that enjoy at least some moral status "will probably prove too overwhelming for any system of norms (legal or moral) to constrain" (p. 2046). Another scholar even goes so far as to note that "researchers will not be happy building anything less than a fully functional synthetic human robot. It is just not in their nature" (Duffy, 2003, p. 188). The point is that robots are already and will likely continue to be designed in ways Bryson would consider "unjust," so the point is moot. Instead of trying to beat back against this trend, scholarly effort would be better directed at dealing with the ethical and legal consequences of the inexorable march forward in robot technology.

5 Bryson has since walked back her use of the term *slaves* because of its association with the inhumanity of slavery throughout human history. See Bryson (2015).

6 See the discussion of relational approaches to robot rights later in this section.

7 I return to these more distant logical connections in Chapter Two.

8 This situation is known as the "problem of other minds" (Dennett, 1981, p. 173). For a review of different theories of mind in the context of social robots, see Gallagher (2013).

9 In this same work, Coeckelbergh explicitly rejects embodiment, although he uses this term in a way that is fundamentally different from Jones' (2013) interpretation. For Coeckelbergh (2011), "[e]mbodiment relations refer to the amplification of bodily perception: technology comes to be experienced as being part of us; we do no longer notice it" (p. 198).

10 For a review of the question concerning moral status as a matter of degree, see DeGrazia (2008).

11 Ecocentrists would not agree that their ethical perspective emphasizes individual entities within the ecosystem. For an extended discussion on ecocentric environmental ethics, see Chapter Four.

12 Given that the present overview focuses exclusively on the writings of scholars who have sought to situate robots in the moral universe, thus far we have only observed examples of anthropocentric and infocentric perspectives. Biocentric and ecocentric perspectives are entertained more fully in the discussions surrounding the rights of nature in Chapter Four.

References

Asimov, I. (1942, March). Runaround. *Astounding Science Fiction*, 94–103).

Birhane, A., & van Dijk, J. (2020). Robot Rights? Let's Talk About Human Welfare Instead. *2020 AAAI/ACM Conference on AI, Ethics, and Society* (pp. 1–7).

Boston Dynamics (2016, February 23). *Atlas, the Next Generation [Video]*. YouTube. Retrieved from https://www.youtube.com/watch?v=rVlhMGQgDkY.

Brown, S. (2013). *Sometimes I Forget You're a Robot*. Dial Books for Young Readers.

Bryson, J. J. (2010). Robots Should Be Slaves. In Y. Wilks (Ed.), *Close Engagements with Artificial Companions: Key Social, Psychological, Ethical and Design Issues* (Vol. 8, pp. 63–74). John Benjamins Publishing Company.

Bryson, J. J. (2015, October 4). Clones Should NOT Be Slaves. *Adventures in NI*. Retrieved from https://joanna-bryson.blogspot.com/2015/10/clones-should-not-be-slaves.html.

Bryson, J. J. (2017, January 31). If Robots Ever Need Rights We'll Have Designed Them Unjustly. *Adventures in NI*. Retrieved from https://joanna-bryson.blogspot.com/2017/01/if-robots-ever-need-rights-well-have.html.

Bryson, J. J. (2018). Patiency Is Not a Virtue: The Design of Intelligent Systems and Systems of Ethics. *Ethics and Information Technology, 20*(1), 15–26.

Bryson, J. J., Diamantis, M. E., & Grant, T. D. (2017). Of, for, and by the People: The Legal Lacuna of Synthetic Persons. *Artificial Intelligence and Law, 25*(3), 273–291.

Chao, B. C. (2010). On Rights and Robots. *Dialogue, 52*(2–3), 97–102.

Coeckelbergh, M. (2010). Robot Rights? Towards a Social-Relational Justification of Moral Consideration. *Ethics and Information Technology, 12*(3), 209–221.

Coeckelbergh, M. (2011). Humans, Animals, and Robots: A Phenomenological Approach to Human–Robot Relations. *International Journal of Social Robotics, 3*(2), 197–204.

Coeckelbergh, M. (2012). *Growing Moral Relations: Critique of Moral Status Ascription.* Palgrave Macmillan.

Coeckelbergh, M. (2014). The Moral Standing of Machines: Towards a Relational and Non-Cartesian Moral Hermeneutics. *Philosophy and Technology, 27*(1), 61–77.

Danaher, J. (2020). Welcoming Robots into the Moral Circle: A Defence of Ethical Behaviourism. *Science and Engineering Ethics, 26*, 2023–2049.

Darling, K. (2016). Extending Legal Protection to Social Robots: The Effects of Anthropomorphism, Empathy, and Violent Behavior Towards Robotic Objects. In R. Calo, A. M. Froomkin, & I. Kerr (Eds.), *Robot Law* (pp. 213–232). Edward Elgar.

DeGrazia, D. (2008). Moral Status as a Matter of Degree? *The Southern Journal of Philosophy, 46*(2), 181–198.

Dennett, D. C. (1981). *Brainstorms: Philosophical Essays of Mind and Psychology* (1st ed.). MIT Press.

Duffy, B. R. (2003). Anthropomorphism and the Social Robot. *Robotics and Autonomous Systems, 42*(3), 177–190.

Enemark, C. (2013). *Armed Drones and the Ethics of War: Military Virtue in a Post-Heroic Age*. Routledge.

Floridi, L. (1999). Information Ethics: On the Philosophical Foundation of Computer Ethics. *Ethics and Information Technology, 1*(1), 33–56.

Floridi, L. (2008). Information Ethics: Its Nature and Scope. In J. Van Den Hoven & J. Weckert (Eds.), *Information Technology and Moral Philosophy* (pp. 40–65). Cambridge University Press.

Fox, W. (2006). *A Theory of General Ethics: Human Relationships, Nature, and the Built Environment*. MIT Press.

Gallagher, S. (2013). You and I, Robot. *AI and Society, 28*(4), 455–460.

Gunkel, D. J. (2012). *The Machine Question: Critical Perspectives on AI, Robots, and Ethics*. MIT Press.

Gunkel, D. J. (2018a). *Robot Rights*. MIT Press.

Gunkel, D. J. (2018b). The Other Question: Can and Should Robots Have Rights? *Ethics and Information Technology, 20*(2), 87–99.

Hubbard, F. P. (2011). "Do Androids Dream?": Personhood and Intelligent Artifacts. *Temple Law Review, 83*, 405–474.

Hume, D. (1980). *A Treatise of Human Nature*. Oxford University Press.

Jonas, H. (1984). *The Imperative of Responsibility: In Search of an Ethics for the Technological Age*. University of Chicago Press.

Jones, R. A. (2013). Relationalism through Social Robotics. *Journal for the Theory of Social Behaviour, 43*(4), 405–424.

Levy, D. (2009). The Ethical Treatment of Artificially Conscious Robots. *International Journal of Social Robotics, 1*(3), 209–216.

Marx, J., & Tiefensee, C. (2015). Of Animals, Robots and Men. *Historical Social Research/ Historische Sozialforschung, 40*(4), 70–91.

Mathews, F. (1991). *The Ecological Self*. Routledge.

McGrath, J. F. (2011). Robots, Rights, and Religion. In J. F. McGrath (Ed.), *Religion and Science Fiction* (pp. 118–153). Pickwick.

Metz, T. (2012). An African Theory of Moral Status: A Relational Alternative to Individualism and Holism. *Ethical Theory and Moral Practice, 15*(3), 387–402.

Miller, L. F. (2015). Granting Automata Human Rights: Challenge to a Basis of Full-Rights Privilege. *Human Rights Review, 16*(4), 369–391.

Nass, C., & Moon, Y. (2000). Machines and Mindlessness: Social Responses to Computers. *Journal of Social Issues, 56*(1), 81–103.

Philippopoulos-Mihalopoulos, A. (2011). Towards a Critical Environmental Law. In A. Philippopoulos-Mihalopoulos (Ed.), *Law and Ecology: New Environmental Foundations* (pp. 18–38). Routledge.

Philippopoulos-Mihalopoulos, A. (2017). Critical Environmental Law in the Anthropocene. In L. J. Kotzé (Ed.), *Environmental Law and Governance for the Anthropocene* (pp. 117–136). Hart Publishing.

Reeves, B., & Nass, C. I. (1996). *The Media Equation: How People Treat Computers, Television, and New Media Like Real People and Places*. Cambridge University Press.

Riek, L. D., & Howard, D. (2014). *A Code of Ethics for the Human–Robot Interaction Profession*. We Robot 2014. University of Miami.

Sachs, B. (2011). The Status of Moral Status. *Pacific Philosophical Quarterly, 92*(1), 87–104.

Serholt, S., Barendregt, W., Vasalou, A., Alves-Oliveira, P., Jones, A., Petisca, S., & Paiva, A. (2017). The Case of Classroom Robots: Teachers' Deliberations on the Ethical Tensions. *AI and Society, 32*(4), 613–631.

Sharkey, A., & Sharkey, N. (2012). Granny and the Robots: Ethical Issues in Robot Care for the Elderly. *Ethics and Information Technology, 14*(1), 27–40.

Solaiman, S. M. (2017). Legal Personality of Robots, Corporations, Idols and Chimpanzees: A Quest for Legitimacy. *Artificial Intelligence and Law, 25*(2), 155–179.

Søraker, J. H. (2007). The Moral Status of Information and Information Technology: A Relational Theory of Moral Status. In S. Hongladarom & C. Ess (Eds.), *Information Technology Ethics: Cultural Perspectives* (pp. 1–19). Idea Group.

Steinert, S. (2014). The Five Robots—A Taxonomy for Roboethics. *International Journal of Social Robotics, 6*(2), 249–260.

Sullins, J. P. (2012). Robots, Love, and Sex: The Ethics of Building a Love Machine. *IEEE Transactions on Affective Computing, 3*(4), 398–409.

Tavani, H. T. (2018). Can Social Robots Qualify for Moral Consideration? Reframing the Question about Robot Rights. *Information, 9*(4), 1–16.

Torrance, S. (2013). Artificial Agents and the Expanding Ethical Circle. *AI and Society, 28*(4), 399–414.

Turner, J. (2019). *Robot Rules: Regulating Artificial Intelligence*. Palgrave Macmillan.

Vermeylen, S. (2017). Materiality and the Ontological Turn in the Anthropocene: Establishing a Dialogue Between Law, Anthropology and Eco-Philosophy. In L. J.

Kotzé (Ed.), *Environmental Law and Governance for the Anthropocene* (pp. 137–162). Hart Publishing.

Veruggio, G., & Operto, F. (2006). Roboethics: A Bottom-Up Interdisciplinary Discourse in the Field of Applied Ethics in Robotics. *International Review of Information Ethics*, 6, 2–8.

Warren, M. A. (1997). *Moral Status: Obligations to Persons and Other Living Things*. Oxford University Press.

2 Getting to rights

Personhoods, statuses, and incidents

'[P]erson' signifies what law makes it signify.

(John Dewey, 1926, p. 655)

Assessing whether or not robots might be eligible for rights requires unpacking and connecting concepts pertinent to the determination of rights. In this chapter, I attempt to accomplish this task by advancing two arguments. First, the debate over the machine question and the discussion of rights for nonhuman entities more generally has suffered from terminological inconsistency and the application of different standards. In particular, participants in these discourses shift between moral and legal frames without fully appreciating how they differ in terms of the criteria applied and the conclusions they reach as a result. Second, returning to Hume for a moment, the *is/ought* problem sets up a false dichotomy between the properties and relational responses to the machine question. Different types of properties connect to different types of personhood, and properties that facilitate interaction between entities cannot be divorced from these relations. Importantly, properties themselves can be relational in nature, and certain mechanisms by which humans identify (with) nonhuman entities reflect a relational kind of personhood. The question pertinent to this inquiry is not whether *is* should come before *ought* or vice versa, but rather, what kind of personhood is under scrutiny? The chapter proceeds by distinguishing between and exploring the relationships among different types of personhood, drawing together personhoods and statuses (i.e., agent/patient and subject/object), explaining how statuses translate into incidents, and finally arriving at theories underpinning the extension of rights. The goal of this chapter is to map the muddled terrain of personhood(s) in the service of clarifying how the concept relates to the different kinds of rights at issue in philosophical and legal scholarship on intelligent machines.

Distinguishing among personhoods: Moral, psychological, legal, and relational

One of the main difficulties experienced in the course of trying to determine whether or not an entity is eligible for rights involves assessing the extent to which it qualifies as a person. This task has proved troublesome for two reasons. First, the definition of a person has varied over time and space. Second, scholars representing different academic fields have tended to talk past one another, neglecting the ways in which related debates over personhood might contribute to greater transdisciplinary knowledge and meaning-making. Many of the contemporary discussions about personhood in law and philosophy begin with definitions put forth by Enlightenment thinkers or their intellectual progeny. A common starting place for locating a widely accepted early definition of a person is the work of John Locke. In *An Essay Concerning Human Understanding*, Locke (1836) describes "person" as a "forensic term" that "belongs only to intelligent agents capable of a law, and happiness and misery" (p. 234). Later, Gray (1909) distinguishes between the common and technical legal meanings of a person, finding that the latter refers to "a subject of legal rights and duties" (p. 27). While many philosophers and legal scholars move swiftly (and perhaps unwittingly) in their writings from the concept of a person to the notion of personhood, anthropologists have spilled considerable ink illuminating the complexities of the latter. Although many different definitions have been proposed throughout anthropology's history (Appell-Warren, 2014), personhood as it is presently understood in the field might be fairly described as a process by which "bodies and persons are culturally conceptualized" (Jackson, 2019, p. 31).

Perhaps unsurprisingly from a practical standpoint, ancient and/or non-Western ideas about personhood have received little attention. After all, what is the value of invoking conceptualizations that fall outside the dominant paradigm when seeking to resolve tensions within it? As I argue in this chapter and throughout the book, this approach has shown itself to be stale and limiting. Considering how other cultures have interpreted personhood is not only theoretically productive, it is normatively crucial to the larger project of recognizing and respecting Indigenous peoples and traditional groups. At least as far back as the 1700s, three kinds of personhood have been identified—moral, psychological, and legal (Vincent, 1989, p. 701).[1] A fourth type of personhood—relational—has long been present in Indigenous and traditional societies, but it gained recognition more recently in the field of anthropology, and has been observed in robot ethics only over the past decade or so. Each of these kinds of personhood are discussed below.

Moral personhood most often evokes a single human who possesses free will, the capacity to act rationally, and self-awareness (Vincent, 1989, p. 701). Gunkel (2012) adds that consciousness has long been considered a "necessary precondition" for moral personhood (p. 90). Himma (2009) defines consciousness as "the capacity for inner subjective experience like that of pain" (p. 19). Scott (1990) identifies the qualities that constitute a person before sketching the contours of moral personhood. Persons are intentional, material, and malleable (i.e., capable of holding an unlimited array of beliefs and belief systems). Moral persons are those who apply the previous traits in the service of fulfilling two classes of needs: (1) those concerned with basic functionalities that enable the development of higher order intentional capacities, and (2) those "needs the meeting of which are necessary for their *continued existence as persons*" (Scott, 1990, p. 80; emphasis in original).[2] Dennett (1976) distinguishes between moral and metaphysical per-sonhood, contending that the latter is a necessary condition of the former. Here, metaphysical personhood includes consciousness, intelligence, and the ability to feel, while moral personhood indicates accountability in the form of rights and responsibilities. Yet, in the end, Dennett (1976) finds that moral and metaphysi-cal personhood "are not separate and distinct concepts but just two different and unstable resting points on the same continuum" (p. 193).

Psychological personhood is also often deployed in the description of individ-ual human beings, but more specifically it entails those capable of demonstrating intentionality,[3] sentience, and self-consciousness, which indicates "awareness of one's own mental processes" (Vincent, 1989, p. 696). Unlike its moral variant, psychological personhood is an empirical phenomenon capturing both deliber-ate and non-deliberate behaviors. It also considers the integration of mental and physical attributes. Perhaps confusingly, Dennett (1976) suggests that psycho-logical qualities are necessary for the assignment of moral personhood (p. 177). For present purposes, it might therefore be useful as a point of distinction to assert that psychological personhood is prior to and works in furtherance of moral per-sonhood, as the former includes the more sophisticated mental processes required to perform the latter.

The concept of (self-)consciousness deserves extended discussion in light of its importance to both moral/metaphysical and psychological personhood. Importantly, attempts to define and operationalize consciousness have been fraught with difficulty. As Gunkel (2012) notes, we still don't really know what consciousness is, and we don't have a surefire way of identifying it in others (p. 90). Similar sentiments were echoed in my interviews with roboticists, each of whom spoke about the lack of a common definition or our poor understanding of the phenomenon. Watson (1979) proposes two avenues through which one might determine whether or not an entity exhibits self-consciousness. First, he suggests a test of introspection: "If you understand what Descartes means when he says, 'I think, therefore I am,' then you have experience of your self which is self-consciousness" (p. 125). Second, we can observe the behavior of others and assess the extent to which it mirrors our own as humans. If they appear to possess the power of communication, they are likely self-conscious.

The debate over consciousness relates back to the theories of mind discussed in the introduction to this book. Under a functionalist theory, cognition is merely a process of inputs and outputs that could potentially be replicated in non-living forms. By contrast, the theory of biological naturalism posits that neurological processes possess a causal power beyond that which inheres in a structure that organizes inputs and outputs. Advocates of this perspective argue that conscious-ness is evidenced through the actual production of mental states, not just fol-lowing the process that led to them. This requires fully duplicating the causal sequences enacted by biological structures, not just simulating them formally through artificial means (Searle, 2008). An alternative view submits that instead of construing consciousness as the product of internal computational processes or cognitive structures, it refers to "the way in which the causal structure of the body of the agent is causally entangled with a world of physical causes" (Manzotti & Jeschke, 2016, p. 172). In this perspective, experience is not wholly an internal phenomenon; the external environment shapes causal outcomes.

Further, consciousness might not be a dichotomous, all-or-nothing quality. Turner (2019, pp. 152–153) maintains that there are in fact at least three dimensions of consciousness. First, there is the kind of consciousness present within a living organism, which ranges from a state of minimal consciousness (i.e., in the midst of REM sleep) to a state of full consciousness (i.e., being wide awake). Second, consciousness may develop over the life of a living being of any species, which implies that newborn babies possess less capacity for consciousness than does a fully mature adult. Finally, varying levels of consciousness may be present across species. While these dimensions are not exhaustive, they do suggest that a purely binary assessment of consciousness is likely to be under-inclusive.

The epistemological challenge of identifying consciousness has led to questions regarding the potential for its presence or absence in artificial forms and resulting conclusions about personhood for nonhuman entities. Setting aside the circular criterion that a person must be a human, rationality would appear to underlie many of the facets of personhood, including consciousness (Pollock, 1989, pp. 111–112). If the appearance of rationality suggests consciousness, and there is no more sophisticated way of establishing that an entity possesses a rational architecture than simply observing it in action, then in principle there is no reason why a machine that acts in ways deemed sufficiently rational could not qualify as a person. Lacking the ability to empirically verify the existence of consciousness, all we have are external interpretations of internal states and observations about the environment in which entities act. If we assume that no one interpretation is any more valid than another, we invite a kind of dynamic subjectivity that permits variable determinations about personhood.

Under modern law, legal consideration only extends to those entities recognized as persons (Donnelly & Whelan, 2018, p. 25). While the Greeks and Romans established enduring boundaries separating property from legal persons (Calverley, 2008, p. 525), since the 18th century Western legal systems have maintained a fairly consistent distinction between natural and artificial persons under the concept of legal personhood. Natural persons typically refer to "(1) human beings, (2) who have been born, (3) who are currently alive, and (4) who are sentient," and in order to exhibit "active legal personality," a person must also possess "sufficient rationality and age" (Kurki, 2017, pp. 75, 76).[4] Artificial persons usually refer to corporations. For some, the dual-pronged model of legal personhood is sufficiently inclusive. Any entities that are not natural persons can simply be considered corporations. Recognizing a third type of legal person "would only raise additional issues" and "create legal uncertainty with no corollary benefit" (Welters, 2013, p. 447).

Echoing Gray, Wise (2010) argues that a person in the legal context needs to satisfy only a single criterion: "the capacity to possess at least one legal right" (p. 1). However, this interpretation ignores important differences between natural and artificial persons concerning their legal function, as discussed below in the context of corporations and ships. One scholar stipulates that the following attributes must be present in order for an entity to qualify for legal personhood: "(1) a person shall be capable of being a subject of law; (2) being a legal subject entails the ability to exercise rights and to perform duties; and (3) the enjoyment of rights needs to exercise awareness and choice" (Solaiman, 2017, p. 161). While rights and duties will be discussed in greater detail later in this chapter, for the time being it suffices to note that the characteristics enumerated above suggest that legal personhood implicitly involves autonomy, intelligence, and intentionality. One must possess a requisite amount of these three properties in order to enjoy rights and perform duties. But Solaiman's (2017) criteria uncritically lump together two distinct, although related, concepts—legal capacity and legal competence. Legal *capacity* involves the ability of an entity to possess rights and discharge duties, while legal *competence* speaks to the ability to enter into legal relations with other entities, which is often dependent upon one's age and level of cognitive functioning (Kurki, 2017, p. 76).

The literature might be characterized as advancing three interrelated conceptions of legal personhood: (1) legal-persons-as-right-holders; (2) capacity-for-rights; and (3) capacity-for-legal-relations (Kurki, 2017, pp. 77–78). The extent to which an entity can *hold* rights depends on the theory of rights used (i.e., will theory or interest theory) and the resulting conclusion obtained from its application. I return to rights theories later. The question regarding whether or not an entity possesses the *capacity* for rights is determined on either a conceptual or legal basis. A better way to state this inquiry might be, is the lack of capacity due to psychological/physical limitations or the entity's mere absence of formal legal recognition? Relatedly, can the dearth of capacity be overcome simply by granting the entity legal status so that it may exercise its rights and fulfill duties, or are there meaningful deficits among the ontological properties of the entity that frustrate its full participation in the legal system? As Kurki (2017) notes in concrete terms, the rock lacks conceptual capacity, whereas the slave lacks legal capacity (p. 83). Finally, an entity's potential for engaging in legal *relations* refers to its ability to partake in the range of reciprocal legal activities specified by Hohfeld (1913), which include rights and other correlative incidents. Hohfeldian incidents are described more thoroughly in a subsequent section. Generally speaking, anything that holds rights can participate in legal relations, but the converse is not necessarily true.

Two specific classes of entities—corporations and ships—are often highlighted in the literature on legal personhood. These usual suspects enter the frame both in situations where authors seek to explain the limited conditions under which non-human entities enjoy legal status and where others argue that the scope of legal personhood can be expanded even further. Here they receive additional consideration with a view towards the latter effort. Although the legal status of corporations was thrust into America's national spotlight in the wake of the U.S. Supreme Court's 2010 decision in *Citizens United v. Federal Election Commission*,[5] jurisprudence and legal theory regarding corporate personhood have engaged in fruitful dialogue since at least 1890 (Matambanadzo, 2013, p. 461).

Four theories allege to establish a basis for the legal personhood of corporations: association theory, grant theory, unique entity theory, and reality theory. In association (or aggregate) theory, a corporation is considered an entity that represents a group of natural persons with a common interest in the operations of a business. In this sense, the corporation acts as a placeholder for real people, limiting their individual responsibility in light of actions undertaken by the larger entity of which they are members. In grant (or fiction) theory, a corporation is recognized as wholly brought into existence by the state for the purpose of promoting the welfare of its citizens. Under this approach, any duties, powers, or rights held by the corporation are expressly contained within its charter, which must be approved by the state. In unique entity theory, a corporation is neither an umbrella organization representing a group of people with common interests nor an organizational artefact created by the state. Instead, it is an entity distinguishable from both natural persons and the state (Kens, 2015, p. 10). Finally, in reality theory, a corporation is a sociological person that exists prior to its formal recognition by law, which serves to institutionalize but not establish its presence. Whereas grant theory treats the corporation as a *de jure* person, reality theory finds that the corporation exists as a *de facto* person (French, 1979, pp. 209–210). Although moral personhood is seen by some as a prerequisite for legal personhood (Koops et al., 2010, p. 548), corporations qualify for the latter without necessarily achieving the former (Solum, 1992, p. 1248).

In the U.S. context, Kens (2015) maintains that not only has the Supreme Court vacillated between theories of corporate personhood, but also that the birth of the legal notion of corporation-as-person stems from the questionable interpretation of language found in seminal precedent on the subject. In *Santa Clara County v. Southern Pacific Railroad*,[6] the complainant challenged a provision of the California Constitution regarding how property values were to be assessed for taxation purposes, arguing that it was unconstitutional to treat railroads differently from other kinds of property. In its decision, the Supreme Court found in favor of the railroad, relying on a technical issue within California law and ignoring the argument advanced in lower courts that the 14th Amendment's equal protection clause applied to corporations and natural persons alike. This relatively benign, if esoteric, ruling might have gone unnoticed were it not for the curious insertion of language by Bancroft Davis, the Supreme Court's reporter. Davis elected to add verbiage from a private memo sent to him by Chief Justice Morrison Waite in which the jurist instructed that

> [t]he court does not wish to hear argument on the question whether the provision in the Fourteenth Amendment to the Constitution, which forbids a State to deny to any person within its jurisdiction the equal protection of the laws, applies to these corporations. We are all of opinion that it does.[7]

Since this case was decided, scholars have debated the validity of this language, with critics arguing that it is not legitimate precedent or that it is perhaps even part of a conspiracy (Kens, 2015, p. 6). At the very least, it has cast a long shadow over the legal basis for corporate personhood among scholarly circles within the United States.

Ships represent another nonhuman entity often invoked in debates about legal personhood. Like corporations, sea-going vessels took a circuitous route to gain the status of legal persons. According to famed American jurist Oliver Wendell Holmes, Jr. (1881), the practice of treating ships, slaves, and animals as legal persons under common law emerged from the human desire for vengeance. The underlying idea was that an injured party needed a way of being compensated for harms suffered under circumstances in which the owner of the proximate cause of the injury was not herself directly culpable. The origins of liability lie in antiquity. Holmes cites the Old Testament and Greek, Roman, and Germanic law as sources of inspiration for what would later become liability doctrine. As societies began to organize themselves using systems of law, the concept of liability was invented to provide a means of resolving this human tendency and obtaining justice for injured parties.

Liability involving vessels constitutes a special case. Ships have been considered a valid subject of liability since at least the Middle Ages (Holmes, Jr., 1881, p. 30). Ships, "the most living of inanimate things" (Holmes, Jr., 1881, p. 26), were tangible assets that could be seized by one's home country in the event of a legal dispute. Their capture could thus serve as an immediate form of remedy where the offending party is foreign to the conflict. The legal personality of a ship was therefore derived "from the compelling fact that it sails the seas between different jurisdictions" (Smith, 1928, p. 288).

The area of admiralty law in the United States evolved a principle designed to redress the grievances of a wronged party without finding the owner of a vessel personally liable—*in rem* proceedings.[8] Present in U.S. law at least as far back as the early 19th century,[9] such proceedings were intended to provide "a form of action pursued to enforce a maritime lien" (Lind, 2009, p. 45). Judgments rendered under the *in rem* principle "affect persons by determining their right to or interest in property. This is the limit of their effect on persons however; they cannot subject anyone to a personal liability, not even for costs" (Fraser Jr., 1948, p. 46). For legal purposes, the vessel, not its owner or captain, is treated as the offender. However, *in rem* should not be understood as dealing with rights that one claims against a thing (as opposed to a person). As Hohfeld (1917) insists, legal relations can only exist between natural persons (p. 721). In *United States v. Brig Malek Adhel*,[10] Justice Joseph Story (quoting Justice John Marshall in *United States v. The Schooner Little Charles*) summarized the logic of *in rem* thusly:

> This is not a proceeding against the owner; it is a proceeding against the vessel for an offense committed by the vessel; which is not the less an offense, and does not the less subject her to forfeiture because it was committed without the authority and against the will of the owner. It is true that inanimate matter can commit no offense. But this body is animated and put in action by the crew, who are guided by the master. The vessel acts and speaks by the

master. She reports herself by the master. It is therefore not unreasonable that the vessel should be affected by this report.[11]

Justice Story's analogical reasoning used to explain the legal status of the brig *Malek Adhel* reflects the fact that "[a]mong inorganic artifacts, none exceeds the sailing ship as an object anthropomorphized in Western civilization" (Lind, 2009, p. 43). British maritime customs had long imbued ships with personality, as sailors referred to their vessels using anthropomorphic language and gave them female names (Mawani, 2018, p. 307). The act of anthropomorphizing ships is thus a ritual of identification that crept its way into common-law systems. However, as demonstrated above, the judges who applied principles that construed seafaring vessels as legal persons did so on the basis of practical expediency, not literal personification derived from religious or cultural beliefs about the ontological status of ships. Ships, therefore, are personified culturally but are determined to possess liability legally. They do not represent the interests of a group, although they have been legalized in ways that shield individuals from responsibility. They are not brought into being purely through state action, although they are recognized as a legal entity for the purposes of engaging in legal relations. Finally, while they are somewhat akin to pre-existing sociological persons, claims of their metaphysical or moral personhood do not form the basis for establishing their legal personhood, which instead relies on a pragmatic approach to resolving legal disputes involving ships. As such, the extension of legal personhood to vessels appears most closely aligned with the unique entity theory of corporate personhood given the fact that ships are neither natural persons nor creations of the state.

Increasingly, legal scholars have turned their attention to the application of legal personhood to technological entities such as artificial intelligence (AI) and robots. The main arguments in this more recent discussion largely fall along anthropocentric lines; that is, they center around the extent to which forms of technology exhibit human-like capabilities or advance human objectives. Although the legal personhood of AI was contemplated by Lehman-Wilzig (1981) almost 40 years ago, Solum (1992) is commonly credited with having articulated one of the earliest and most extensive deliberations on the subject.[12] His analysis raises three objections to affording AI legal personhood: (1) personhood should only be granted to natural persons (i.e., humans); (2) AI lacks some property necessary to qualify for legal personhood (i.e., consciousness, intentionality, a soul, etc.); and (3) human artefacts can never amount to more than human property (Solum, 1992, p. 1258). In the end, he surmises that the question of legal personhood will be answered through an improved understanding of how the human mind works (a neurological response) and our experience with AI (a phenomenological response).

Calverley (2008) adds that intelligent machines could overcome the gulf between property and legal person as long as they exhibited "a level of mental activity in areas deemed relevant to law, such as autonomy and intentionality" (p. 527). The author concludes that demonstrating the functional equivalent of intentionality is "probably enough" to bestow legal personhood on technological entities (Calverley, 2008, p. 534). For Koops et al. (2010), the possibility of assigning legal personhood to technological entities will depend on the method by which the law is approached (i.e., functionalism, legal positivism, naturalism, etc.) and the level of sophistication exhibited by the entity in question. This leads to three overlapping evolutionary stages of legal adaptation—"short term: interpretation and extension of existing law," "middle term: limited personhood with strict liability," and "long term: full personhood with 'posthuman' rights" (Koops et al., 2010, pp. 554–559). Ultimately, humans will have to decide whether the entities in question are merely the products of human ingenuity or intelligent machines capable of their own responsibility and will (Andrade et al., 2007, as cited in Koops et al., 2010, pp. 560–561, n. 220).

Hubbard (2011) proposes a behavioral test in which an entity might qualify for personhood if it can demonstrate all of the following:

> (1) the ability to interact with its environment and to engage in complex thought and communication, (2) a sense of being a self with a concern for achieving its plan of or purpose in life, and (3) the ability to live in a community based on mutual self-interest with other persons.
>
> (p. 419)

These capacities ultimately depend on the possession of properties such as rationality, intelligence, (self-)consciousness, and emotions. Putting forth the kind of practical reasoning observed in jurisprudence on ships, Pietrzykowski (2017) argues that AI might be worthy of legal personhood if granting that status would assist humans in determining liability for actions taken by artificial agents. Solaiman (2017) offers that robots are artefacts that do not (currently) possess sufficient autonomy to render their actions self-controlled, which means they cannot enjoy rights or fulfill duties, thus precluding them from being deemed proper legal subjects. Corporations, by contrast, consist of people; religious idols, though not human, have interests tended to by people. In short, the degree of separation from humans dictates whether or not an entity might quality for legal personhood.

Despite the various approaches to legal personhood for corporations, ships, or technological entities surveyed here, the literature tends to reflect a human-centered view that is nearly a century old:

> the function of legal personality … is to regulate behavior, it is not alone to regulate the conduct of the subject on which it is conferred; it is to regulate also the conduct of human beings toward the subject or toward each other.
>
> (Smith, 1928, p. 296)

Yet, at least some analysts have shown a preference for more complex notions of legal personhood. Koops et al. (2010) identify concentric classes of persons in order of increasing personality, ranging from abstract or virtual entities in the outermost ring to abstract or virtual persons, legal persons, moral persons, and finally social persons in the innermost ring. Under this configuration, virtual entities such as AI could potentially constitute legal persons to the extent that they are attached to certain legal rights or duties. However, only an entity that possesses both legal capacity *and* legal competence would be considered a true legal person. Kurki (2017) goes a step further, suggesting that legal personhood might be better understood as a "cluster concept" that leaves indeterminate the line between legal personality and nonpersonality (p. 84). Tasioulas (2019) suggests that the kind of legal personhood bestowed upon robots and AI need not be the same as that which humans enjoy. Instead, intelligent machines might qualify for an attenuated form of legal personhood attached to more modest bundles of rights and responsibilities that vary according to the type of technology under scrutiny. The multi-spectral framework presented in Chapter Five draws inspiration from some of the above ideas in the hopes of constructing a complex and contingent conceptualization of personhood that acknowledges its historical plasticity.

While the preceding three types of personhood rely on the properties of the entity in question, a fourth kind of personhood focuses on the relations among entities. A relational personhood ("one among others") is one in which interpersonal relationships constitute identities (Splitter, 2015, p. 2). Although the discussion of relational personhood has spanned several disciplines, it owes a particular intellectual debt to anthropology, which has engaged in a productive conversation about the differences between Western and non-Western cultural perspectives on persons and personhood since at least as far back as the 1930s (Appell-Warren, 2014, p. 33). In a widely cited article on animism, or the tendency to recognize the lifelike quality of things through our relations with them, Bird-David (1999) defends the relational epistemologies of traditional peoples, which entail "knowing the world by focusing primarily on relatedness, from a related point of view, within the shifting horizons of the related viewer" (p. S69). Such non-Western approaches to understanding the world offer alternative views on the kinds of beings present in social contexts and how we should navigate our interactions with them.

Although nonhuman entities may not constitute persons according to the standards of Western empiricism, their treatment by non-Western societies emerges from an alternative vision of person-*hood*.[13] For example, a study of the ways in which Classic Mayan people engage with objects suggests that "personhood fundamentally does not require humans as a source, acting instead as an untethered resource that is accessed by entities (human or not) that are able to act in social, relational ways" (Jackson, 2019, p. 32). With a nod towards the pioneering work of Bird-David, Fowler (2018) argues that "personhood is always relational" (p. 397). Given the plurality of ways in which personhood is conceived by different groups and the qualitatively variable nature and strength of relationships, Fowler proposes a dynamic framework for assessing personhood that eschews static determinations of properties in favor of examining tensions along four different dimensions or axes of relationality—fixed/mutable, independent/interdependent, bounded/distributed, and typical/distinctive. This flexible heuristic device captures the variation and complexity found in different conceptions of personhood

while paying specific tribute to the relational emphasis identified in the epistemologies of traditional cultures.

In what some have described as a "relational turn" in the study of robotics
(Jones, 2013, p. 405), literature on the moral status of intelligent machines has
drawn inspiration from the concept of relational personhood found in anthropology.[14] Søraker (2007) puts forth a "relational theory of moral status" that consists
of a Western-derived intrinsic component and a relational component based on
East Asian philosophy (p. 2). The intrinsic aspect relies on the extent to which an
entity possesses capabilities or merely abilities (essentially different groupings
of properties), arriving at a hierarchically organized system of tiers ranging from
"merely sentient beings" (lowest level) to "merely self-conscious beings" (middle
level) to "moral persons" (apex level) (Søraker, 2007, p. 5). The relational aspect
depends on the irreplaceability and constitutivity of an entity and the degree to
which possession of those qualities contributes to one's practical identity. The
result is a fourth tier added just below "merely sentient beings," which is termed
"non-sentient entities" (Søraker, 2007, p. 15). This final tier is reserved for those
entities determined to be irreplaceable to and constitutive of a person's practical
identity, which afford them moral status but not moral standing.

Forms of technology could conceivably satisfy the criteria required for this
fourth category. For instance, were an intelligent machine to be considered an
irreplaceable companion whose significance makes a person feel as though life is
worth living, it might be a non-sentient entity worthy of moral status. Two examples from science fiction come to mind—the romantic relationship established
between Theodore Twombly and his AI love interest Samantha in the film *Her*
(2013), and the maternal relationship between Mother (a humanoid robot) and
Daughter (a synthetically grown human), who is raised in a repopulation facility
without any human interaction, as depicted in the film *I Am Mother* (2019). In
both of these examples, the combination of intrinsic properties and relational criteria suggest that the technological being in question might qualify for (the lowest
tier of) moral status.

Coeckelbergh (2010) takes an approach similar to Søraker's, largely rejecting
a properties-based (i.e., intrinsic) account of moral consideration in favor of a
social-relational (i.e., extrinsic) one. Drawing from Western ecology and non-
Western worldviews, he proposes a "radically relational ecology" that recognizes
the moral significance of relations among all entities present in a social-ecological
system (Coeckelbergh, 2010, p. 216). The epistemology of this approach shares
with Fowler's anthropological framework for personhood an emphasis on contingency and context, and a reliance on experience and imagination, as opposed to *a
priori* assumptions, as tools useful for determining the status of an entity. As such,
far from using a checklist to separate persons from nonpersons, Coeckelbergh's
(2014) relational account makes moral distinctions on the basis of historical and
phenomenological conditions "entangled with subjectivity" (p. 66).

Gunkel (2018), leaning on the work of Emmanuel Levinas, argues that how we
ought to treat another entity precedes our determinations regarding what it *is* on
the basis of its ontological properties. Importantly, this means that anything with

whom we find ourselves engaging in relations could "be considered a legitimate moral subject" (Gunkel, 2018, p. 167). This approach is not only a philosophical exercise; it also finds support in empirical studies of the relations between humans and machines. The author cites numerous experiments in which humans afford social standing to computers. Although perhaps less specified than Coeckelbergh's account, Gunkel's contemporary application of Levinasian philosophy nevertheless provides intellectual and practical foundations for establishing the personhood of nonhuman mechanical entities by virtue of our relations with them.

While Gunkel relies on Western scholarship, he acknowledges the similarities that his approach shares with ideas found in non-Western cultures. In particular, he references Jones's (2016) work on personhood and social robotics in Asia. Although the space he dedicates to non-Western conceptions of nonhuman entities is relatively brief, it presents a useful jumping-off point for a more extended discussion. One non-Western culture that brings a relational view of personhood to bear on technological entities is Japan. Japan has long held a reputation as a society hospitable to, if not enthusiastically embracing, robots.[15] Its cultural experience with robots extends at least as far back as the 17th century, when specialized machines carried tea for patrons (Vallverdú, 2011, p. 176). Today, Japan is known as the "Robot Kingdom" (*robotto okoku*) (Schodt, 1988).[16]

Many observers have written that the basis for Japan's attitudes towards robots lies in Shinto, the country's Indigenous religion.[17] Aspects of Shintoism inform cultural predispositions towards robots in Japanese society. First, everything in the universe shares the same parents and is thus related. Therefore, Gods (*Kami*), humans, and nature possess a kind of kinship with one another (Herbert, 1967, p. 21). Second, each object or natural being possesses its own spirit or soul (*tama*) (Kitano, 2006, p. 80). This belief has translated into a culture of animism in which even artificial objects not constructed out of purely natural materials, such as robots, are thought to possess their own spirit. On this basis, ostensibly non-living things enjoy "the same ontological status as living entities" (Vallverdú, 2011, p. 178). However, it is only when such objects serve as tools or otherwise perform tasks in harmony with their human owners that they "come alive" (Mitsukuni et al., 1985, p. 90). But as Yueh-Hsuan Weng noted in an interview, unlike Christian societies of the West, Japanese culture allows robots to be more than tools; they can be friends, too.

Importantly, the Japanese system of ethics governing human–object interactions is only activated in the context of human relationships (Kitano, 2006, p. 82). In a sense, natural and artificial objects alike gain a kind of personhood only through their relations with humans. And just like living beings, artificial objects such as robots enjoy a limited time on this Earth. Funerals have even been held to memorialize the "lives" of hundreds of defunct Sony Aibo robotic dogs and return their souls to their rightful owners before their bodies are harvested for mechanical parts (Neuman, 2018). But robots can also outlast their owners. As explained by Atsuo Takanishi in an interview, an intelligent machine left behind in the wake of its owner's demise can serve as a conduit through which grieving humans might communicate with their deceased relative. In short, robots occupy a practical and emotional place in Japanese society that arguably stems from its native religious doctrine.

Generally speaking, both Western and Eastern "ways of worlding" (Blaser, 2014, p. 53) offer mechanisms that suggest that personhood might be seen through a relational lens. They range from the abstract to the concrete, and from anthropocentric to non-anthropocentric. Each perspective described below conveys a means by which an entity is thrust into relations with, and thus rendered meaningful to, an Other. The literature on environmental ethics presents a useful starting point and takes a nod from the science of ecology. Deep ecologist Arne Naess (1995) explains how self-realization (i.e., "the fulfillment of potentials each of us has") results in a "broadening and deepening of the self," which animates our concern for the wellbeing and thriving of other entities (p. 226). As we become more fully realized individuals, we come to identify with others enduring the same struggle. The more we come to see ourselves in others, the more we will be driven to fight for the self-realization of those with whom we identify. The endpoint of self-realization is unity; our interests become inseparable from those of other beings. To be sure, deep ecology does not help determine which entities are the most deserving of personhood. Rather, it specifies the mechanism through which persons come to obtain the same ontological status as their *relata*. Fox's (1990) transpersonal ecology takes Naess's ideas further, proposing cosmological, personal, and ontological forms of identification. While the bases of identification differ across types (i.e., non-Western worldviews, physical or emotional contact, and the notion that all things exist, respectively), the main idea remains consistent—we relate to nonhuman entities by expanding ourselves into them. As such, humans don't locate personhood in other beings so much as they recognize all with which we stand in relation to.[18]

Teubner (2006), although ostensibly discussing legal personhood, asserts that personification is a strategy humans employ to manage uncertainty in situations where the internal properties of another entity are unknown, such as in the case of animals and electronic agents. Drawing on insights from Luhmann and Latour, the author holds that personifying nonhumans leads to their admission into different social contexts of the political ecology on the basis of their possession of variable degrees of agency. Aaltola (2008), writing about animal ethics, contends that one of the strongest contemporary approaches to determining personhood focuses on the capacity of entities to interact with (and thus relate to) others. Although not specifically addressing the issue of personhood, Coeckelbergh (2010, 2011) advocates in favor of a similar approach in the context of moral relations with animals and robots. He concludes that a phenomenological perspective that emphasizes how humans experience a robot's appearance-in-context dictates our relations with it. For Coeckelbergh and Gunkel (2014), the act of naming animals affords them a face that inserts nonhuman beings into moral relations with humans. Darling (2016) argues that the human tendency to anthropomorphize entities by projecting human-like qualities onto them could have legal implications. In particular, social robots might present a compelling case for legal protection by virtue of their physicality, perceived autonomous movement, and capacity to engage in social behavior.

Lewis et al. (2018) discuss how Indigenous worldviews might engage with intelligent machines in the context of kinship, the kind of relations in which "persons participate intrinsically in each other's existence" (Sahlins, 2011, p. 2). For the Cree people, both animate and inanimate things might be welcomed into their "circle of kinship" (*wahkohtowin*), but AI in particular might prove to be a hard case because of its perceived lack of "humanness" or "naturalness" (Lewis et al., 2018, p. 7). For the Lakota, the interiority (i.e., consciousness, intentionality, soul, etc.) of an entity is crucial to its status, although such attributes of internal life may be extended to members of the nonhuman world through their possession of spirits. Because AI is constructed out of earth materials that exhibit interiority due to the spirits contained within them, its agency lies within the natural elements used in its production. In essence, according to Lakota ontology, the ultimate materiality of technology establishes the spiritual link that brings humans and nonhumans into reciprocal relationships with one another. The diversity of perspectives among Indigenous groups briefly described here suggests that the prospects for relational personhood held by intelligent machines might vary greatly across traditional cultures.

The personhoods discussed in this chapter are all crucial to the question of rights for robots. However, most work on this question references only one kind of personhood, neglecting the important ways in which personhoods connect and how these connections affect the conclusions reached. As mentioned above, psychological personhood is often viewed as a precursor to moral personhood, which some have argued determines legal personhood. Others counter that demonstrating psychological personhood directly qualifies an entity for legal personhood. The addition of a fourth kind—relational personhood—renders the picture more complex. As detailed in this section, each of the other three personhood types relies on assumptions that are, I argue, fundamentally relational in nature. The criteria for psychological and moral personhood are properties possessed by humans. As such, entities can qualify for either of these types to the extent that *we*, humans, can relate to *them*.

While scholars have yet to identify the magical combination of properties that conclusively delimits the kinds of entities worthy of psychological or moral personhood, legal personhood has proven, at least in theory, more bounded, though not without controversy. The main argument surrounding legal personhood holds that its extension to nonhuman entities is warranted when doing so helps resolve conflicts between humans. It is not because courts view natural persons and their artificial brethren as literal kin. Legal personhood is "generally accepted as a useful fiction" (Youatt, 2017, p. 43). It is useful to *us*.

Western legal systems evolved in such a way so as to grant corporations and ships admittance to courtrooms full not of their peers, but of humans for whom treating nonhumans as legal persons is expedient, if awkward, idiosyncratic, and historically variable. Therefore, the first three types of personhood are relational in the sense that they reflect human qualities (i.e., the conditions under which we can relate to other entities) and structure relations between humans and nonhumans in political and legal institutions (i.e., how we can relate to other entities in the human-created world). It is through qualities shared or relationships with humans that nonhuman entities gain identities and qualify for one or another type of personhood. Importantly, relational personhood also serves to bridge the gap between Western and non-Western perspectives on personhood. Considering on equal footing the myriad ways in which humans relate to nonhumans raises the possibility of disrupting the dominant rights paradigm with more inclusive models.

Distinguishing among statuses: Moral and legal

Under the conventional approach that moves from determining what an entity *is* to how it *ought* to be treated, an intermediate step along the path from personhood to rights involves assessing an entity's moral or legal status. The presence or absence of certain ontological properties dictates the status of a being in question, with attendant consequences for the range of incidents that might apply to it. In other words, the properties possessed by an entity determine the form of personhood for which it qualifies, which in turn affects whether it can be seen as a moral agent or moral patient, and/or a legal subject or legal object. I argue that relations are also relevant to this analysis. The kinds of properties deemed integral to moral status ascriptions reflect human-like traits that facilitate identification with nonhuman entities, and the capacities central to qualifying for legal status represent how dominant actors in legal systems construct relations with other beings when convenient or normatively appropriate. This section seeks to sketch these ideas in greater detail by unpacking the antecedents of and differences between moral and legal statuses with an eye towards understanding the kinds of benefits or burdens an entity might face.

First, the properties associated with psychological and moral personhood (i.e., consciousness, intelligence, rationality, sentience, etc.) directly influence an entity's *moral* status. Just like with personhood, there is no precise or widely agreed-upon combination of properties that clearly identifies a moral agent or moral patient. For present purposes, more general definitions of these terms will have to suffice. Broadly speaking, "[w]hereas a moral agent is something that has duties or obligations, a moral patient is something owed at least one duty or obligation" (Himma, 2009, p. 21). While moral agents are often also moral patients (i.e., adult humans), the reverse is not necessarily true (i.e., human infants). The main difference lies in the extent to which an actor can be deemed accountable for her actions. To this end, two capacities are associated with moral agency—free will and understanding the difference between right and wrong (Himma, 2009, pp. 22–23).[19] By contrast, moral patiency requires "the capacity to be acted upon in ways that can be evaluated as good or evil" (K. Gray & Wegner, 2009, p. 506). Of course, this definition indicates more about a given society, its social norms, and the kinds of beings it affords moral consideration than it does the characteristics of a particular entity. One need only look at animal-welfare laws around the world to observe that domesticated pets like dogs and cats can qualify as moral patients in many jurisdictions given contemporary human sentiment regarding their fair treatment.

Scholars writing on AI and robot ethics have engaged in a vigorous debate over whether or not technological entities might qualify for moral agency or moral patiency (or both). To begin, Arp (2005) reverses the logical order of the scheme outlined in the previous section, arguing that, *inter alia*, a being needs to be "considered a responsible moral agent" in order to be a person (p. 121).[20] Writing specifically about androids in the science fiction film series *Star Wars*, Arp concludes that R2-D2 and C-3PO possess other traits inherent to personhood—reason/ rationality, mental states, language, and participating in social relationships—but he sidesteps the issue of their moral agency entirely. Johnson (2006) adheres to the Western philosophical account of moral agency, which privileges behavior that exhibits autonomy, intentionality, and responsibility, all of which imply the existence of mental states. But to move from moral agency to action, these internal traits must produce an act rationally directed at a moral patient in the external world. In the end, she finds that while computers might be able to satisfy most of these criteria, their intentionality is ultimately not their own but rather that of programmers, the system, and the user. Therefore, intelligent machines might be more appropriately classified as "moral entities but not alone moral agents" (Johnson, 2006, p. 203; emphasis omitted).

Sullins (2006) urges that robots need to satisfy three requirements in order to be considered moral agents—significant autonomy from operators or programmers, behavior that can only be explained by intentionality, and actions that convey a sense of responsibility towards another moral agent (p. 25). Interestingly, he also maintains that evidence of these behaviors directly qualifies a robot for moral rights whether or not they enjoy personhood. However, as indicated here, the ontological properties that underlie Sullins's behaviors are the very same that determine an entity's psychological or moral personhood. Gunkel (2012) remarks that inquiring about the kind of moral status intelligent machines might possess has largely proven to be a failed enterprise. He adds that this line of investigation has instead cast doubt on the moral agency of humans. Previewing the argument he makes at length in *Robot Rights*, Gunkel observes that endeavoring to define moral agency prior to encountering another entity is a fool's errand. Determinations regarding moral agency arise in the course of interacting and establishing relationships with others. The pertinent issue, as Gunkel sees it, is whether or not nonhuman entities are moral patients to whom we have moral duties and responsibilities.

Marx and Tiefensee (2015) proffer strict guidelines governing the demonstration of moral agency, which hinge on the ability of entities to "be held morally responsible for their actions" (p. 72). This responsibility depends upon an actor's rationality and autonomy. The authors conclude that while current robots are not sufficiently autonomous to qualify as moral agents, future robots operating under strong AI might if they come to understand and act upon moral obligations. More recently, Bryson (2018) refers to a moral agent as "something deemed responsible by a society for its actions" and a moral patient as "something a society deems itself responsible for preserving the well being of" (p. 16). As a skeptic of any effort to elevate the moral status of technology, she maintains a position on the subject that mirrors the pragmatism described in the context of legal personhood for corporations and ships. That is, the only justifications for treating intelligent systems as moral agents would be that doing so would enable greater human control over technology and that the benefits of such AI would outweigh the loss of human moral responsibility. Crucially for Bryson, moral status ascription is a zero-sum game. At best, society could deem intelligent machines second-order moral patients that do not compete with human interests, but avoiding the creation of such advanced technological entities would be the more preferable route.

The survey of extant literature provided here offers tentative support for Gunkel's (2012) objection—theory on moral agency in the context of AI and robots is "confused and messy" (p. 91). The philosophical works reviewed in this section reach a range of conclusions about the moral agency of technological entities, including that it does not apply, should not apply, does not yet apply, might apply, and is less relevant than moral patiency. Most of the scholarship discussed above adheres to a properties-based approach to moral status that leads to a competition whose victor remains elusive. In the face of such indeterminacy, it stands to reason that alternatives like the relational approach should at least be allowed to enter the race. For the moment, the only meaningful conclusion about moral agency/patiency that this non-philosopher can endorse is that there is no settled upon grouping of ontological properties that one can point to in order to justify ascription of one status or the other. However, if we accept that relational personhood precedes its psychological and moral variants, we move in what I argue is a more productive direction—probing the interaction effects between properties and relations to better understand the conditions under which encounters among entities lead to recognition in various contexts.

Second, the equally messy world of legal personhood and its constituent properties informs the determination of an entity's legal status. The confusion begins with the fact that the term "legal subject" (*Rechtssubjekt*) is used in civil law jurisdictions to denote all legal persons, while those writing from the perspective of common law traditions distinguish between natural and artificial persons, as mentioned earlier (Kurki & Pietrzykowski, 2017, p. viii). To make matters even murkier, as shown above, some observers argue that legal personhood itself is ultimately dependent upon the presence of properties associated with psychological and moral personhoods. For example, Solaiman (2017) states that in order to enjoy legal personhood, an entity needs to be capable of being a legal subject. Next, in order to be a legal subject, a being must be able to exercise rights and perform duties. Finally, in order to exercise rights, an entity needs to exhibit awareness and choice. These last two traits reflect the cognitive functioning of the entity in question, and thus relate to properties identified with psychological and moral personhood. As such, Solaiman's notion of legal subject includes Kurki's concept of legal competence.

A seemingly minor but nonetheless important definitional divide centers on whether it is enough that an entity is *capable* of holding rights in order for it to be considered a legal subject (which itself is determined by evidence of conceptual or legal capacity, according to Kurki), or whether it must actively *possess* rights to achieve that legal status. Solaiman resides in the former camp along with Knauer (2003), for whom capacity serves as the linchpin of legal status. She writes that "the determination of *in*capacity represents a crucial dividing line between legal subjects and those who are the object of legal protections" (Knauer, 2003, p. 323; emphasis added). While legal subjects possess legal competence and can be held responsible for their actions, legal objects (i.e., property) lack legal competence and cannot be held responsible for their actions. A member of the opposing camp, Turner (2019) contends that "[a] legal subject is an entity which *holds* rights and obligations in a given system. The status of legal subject is something which is thrust upon a person, animal or thing" (p. 42; emphasis added). Here, it is the possession of rights, not merely the capacity to possess them, that determines the legal status of an entity already found to obtain legal personhood. Turner adds that a "legal agent" is a subtype of legal subject that "can control and change its behaviour and understand the legal consequences of its actions or omissions" (Turner, 2019, p. 43). This subtype evokes the cognitive abilities of an entity, which again speak to properties underlying psychological and moral personhoods.[21]

Setting aside the properties-based debate between capacity and possession of rights, another way of addressing the question of legal status proceeds on relational grounds. Here, what matters are not the characteristics of an entity, but the way in which it relates to other, often more powerful, entities. Ownership and recognition are two avenues through which relationships, and thus legal statuses, gain clarification. In many jurisdictions, if something is owned it is an object, not a legal subject (Cullinan, 2003, p. 77). Objects (not in the purely ontological sense, but in the legal world) are considered property, lacking the agency necessary to participate in legal relations. In its classic legal interpretation, property represents "that sole and despotic dominion which one man claims and exercises over the external things of the world, in total exclusion of the right of any other individual in the universe" (Blackstone, 1766, p. 2). Under this view, the concept of property is laden with imbalanced power relations. A more modern take on property advances a relational perspective defined by obligations people have towards others (Davies, 2007, p. 2). Thus, while the traditional interpretation of property draws a bright line between objects and subjects (affording the latter exclusive rights over the former), a more contemporary assessment at least considers how one's use of property holds implications for those outside of immediate ownership relations, even if it does not disrupt the object–subject distinction.

Recognition offers another means of altering legal status. By deliberately elevating the position of a class of entities within a legal system (i.e., recognizing their legal personhood), new actors emerge to find themselves suddenly instilled with legal competence, if not legal capacity. The progressive extension of legal recognition from land-owning men to marginalized groups such as Indigenous peoples and women serves as a case in point (Vermeylen, 2017, p. 159). This gradual expansion of the legal circle demonstrates the contingent and fictive foundation upon which the concept of legal subject rests. Power dynamics are thus an inescapable part of recognition, as status changes are given effect by those who enjoy a position of authority in legal systems.

In fairness to the analysts discussed in this section, disciplinary silos and appropriately tailored research questions have caused the intellectual quests to explain the basis for moral status and legal status to occur in parallel. I have argued that there is much these pursuits share in common given their emphases on ontological properties. However, the story does not end there. I have also suggested that despite the seemingly separate pathways through which psychological and moral personhoods inform moral status and through which legal personhood dictates legal status, both of these statuses are influenced by mechanisms of relational personhood. The extent to which nonhuman entities appear and behave in ways similar to us and the degree to which they prove capable of performing roles that grant them entrance into institutions designed by us determine the status they hold. While at present the properties-driven criteria used for deciphering moral or legal status seem unsettled at best and unhelpful at worst, a relational approach, though at times anthropocentric in nature, offers a useful alternative for determining these statuses. The next challenge involves demonstrating how these statuses interface with a range of incidents, including our main interest—rights.

Distinguishing among incidents: Moral and legal

Defining rights at times seems like a Sisyphean task. Questions abound as to their origins, referents, applications, and limits. What is the basis for their extension? To whom do they apply? Under what conditions? What do we do when one person's rights conflict with those of another? The difficulty of establishing precise boundaries around the concept of rights is amplified by the fact that the term has been used to describe different relationships among actors in the legal system (Shestack, 1998, p. 203). In an effort to clarify the range of legal relations that may exist among persons beyond the common catch-all categories of rights and duties,[22] Hohfeld (1913) developed a typology outlining a series of opposite and correlative incidents. Pairs of opposites include the following: rights/no-rights, privileges/duties, powers/disabilities, and immunities/liabilities. Correlative legal relations are also divided into four pairings: rights/duties, privileges/no-rights, powers/liabilities, and immunities/disabilities.

The idea of opposite incidents is fairly self-explanatory. If one possesses a(n) right/privilege/power/immunity, the absence of such a benefit would indicate a no-right/duty/disability/liability. Correlatives are a bit more complex. In this latter category, one's possession of a(n) right/privilege/power/immunity implies the reciprocal duty/no-right/liability/disability of another. Of particular interest are rights and duties. The former are derived from the latter, which constitute "obligations to refrain from over-interfering with a party exercising his or her privilege" (Manus, 1998, p. 574). For example, if X has a right to vote, Y has a correlative duty not to impede on X's ability to cast a ballot in an election. Thus, correlatives describe legal relations in which one's enjoyment of a legally sanctioned benefit necessarily imposes restrictions on another as a means of protecting the first person from potential violations committed by the second. Importantly, this scheme permits an examination of legal relations from the perspective of the patient (i.e., the person possessing the right) or the agent (i.e., the person who has a duty not to infringe upon that right) (Gunkel, 2018, p. 29).

Hohfeldian incidents have been the subject of extensive discussion in philosophical and legal circles, but they are mainly descriptive, not prescriptive (Kelch, 1999, p. 8); that is, they do not provide a normative basis for the specific content of legal duties, rights, and so on. Rather, they simply chart the universe of legal relations among human beings, whom the author considered to be the only proper subjects of law (Hohfeld, 1917, p. 721). Fortunately, some writers have taken up the mantle of unpacking and critiquing this typology in order to assess its utility in specific domains of law, especially where nonhuman entities are concerned. Sumner (1987) examines the prospect for animal rights using a Hohfeldian framework. He finds that animals would not qualify for rights under this scheme because they do not possess the capacity to adhere to normative rules. Manus (1998) describes how Hohfeld's incidents apply to environmental matters before the law. He determines that the scheme's usefulness resides in its distinction between rights and privileges. In the United States, the legal presumption is that relations involving the environment are characterized by privileges, not rights. For example, while the privilege of accessing a public beach does not imply a legal right to beach access, others have no right to limit our access to or despoil this natural amenity. However, still others might simultaneously possess the privilege of exploiting nature.[23] Kelch (1999) holds that the Hohfeldian foundation for rights—the existence of correlative duties—denies the plurality of sources from which rights emerge. The author offers an alternative foundation for rights that applies in the specific context of animals. Here, since emotions influence our conception of morality, and morality affects the choices we make regarding the kinds of entities extended legal rights, emotions must affect our determination of rights. Thus, our emotional reactions to animals help us understand their interests, and it is through identifying these interests that we can assess the extent to which animals deserve rights.

Gunkel (2018) reiterates the criticism that Hohfeld's framework "does not explain who has a right or why" (p. 30). He adds that while Hohfeldian incidents were initially conceived to flesh out legal rights, they have been adapted to describe moral and political rights as well (Gunkel, 2018, p. 27). This is a crucial point, as some analysts have equivocated between moral and legal rights without fully acknowledging the different properties, personhoods, and statuses from which they stem. Perhaps the most well-known translation of Hohfeld's work into the domain of morality is Wellman's (1985) *A Theory of Rights*. Here, the author identifies five moral positions that correspond to the legal incidents found in Hohfeld's scheme. Each of these positions has a correlative of sorts, although some involve the possessor while others reflect reciprocal obligations held by second or (in the case of moral sanctions) third parties. The positions and their correlatives (roughly) are as follows: moral duties/sanctions, moral claims/duties, moral liberties/no-duties, moral powers/intentions, and moral immunities/no-abilities.[24] Together, these positions constitute moral rights and explain the roles played by moral agents and moral patients.[25]

As Solum (1992) rightly pointed out nearly three decades ago, determining legal personhood (and by extension legal rights) does not follow the same investigative path used to assess moral status (p. 1240, n. 36). The kinds of legal incidents relevant to parties engaged in legal relations depend on an entity's status as a legal subject/object, which reflects the extent to which it qualifies for legal personhood. The kinds of moral positions that apply to actors in a given situation can be traced back to an entity's status as a moral agent/patient and the extent to which it qualifies for psychological or moral personhood. However, I argue that what legal and moral personhoods have in common are their underlying relational characteristics. Indeed, as Wellman (1985) observes, Hohfeld's legal positions are essentially relational in nature. As mentioned in the previous section, legal and moral personhoods are predicated on the presence of properties associated with humans and the extent to which extending personhood serves human purposes. The legal and moral incidents/positions described in this section represent an extension of these subjective rules. That there are moral positions analogous to the legal ones and that morals are fundamentally relational in the sense that they govern relations among members of society suggests that relationality is a conceptual, though not causal, thread that connects legal and moral personhoods and their attendant incidents/positions.

Having defined the requisite properties/mechanisms, personhoods, statuses, and positions/incidents, and demonstrated the relationships between them, it now becomes possible to construct a visual representation of the concepts relevant to this study and how they interface with each other (see Figure 2.1).

Figure 2.1 conveys how the various concepts described earlier relate to one another. The relationships are associational, not causal, in nature. Connectors featuring arrows depict the directionality of logical sequencing among concepts. Connectors without arrows simply display related concepts whose direction of influence remains undefined. The more complete the line of the connector, the stronger the association between concepts (i.e., solid lines express stronger relationships than dashed lines, which entail relationships stronger than those denoted by dotted lines). For instance, under a conventional approach, an entity that exhibits rationality might qualify for moral personhood, which is a quality possessed prior to determining an entity's status as a moral agent or moral patient. Identifying whether an entity qualifies as a moral agent or a moral patient influences the kinds of moral positions (Wellman's translation of Hohfeldian legal incidents) that apply to it. The far-left side of the image demonstrates how ontological properties differ qualitatively from mechanisms. Whereas ontological properties feed into legal ones and various groups of properties relate to different forms of personhood, mechanisms (fifth and sixth boxes on the bottom left) lead to determinations about the existence of relational personhood—the antecedent to other personhoods. Yet, mechanisms and properties are reflexively related to one another; the directionality of those relationships remains unclear. Then psychological/moral and legal personhoods shuffle into their respective statuses. Finally, the kind of status an entity is determined to possess affects the sorts of moral positions or legal incidents to which it is entitled. To summarize, relational mechanisms active in both

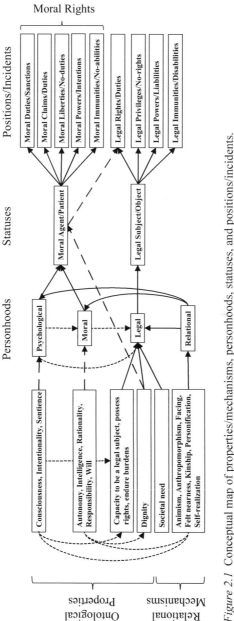

Figure 2.1 Conceptual map of properties/mechanisms, personhoods, statuses, and positions/incidents.

Western and non-Western cultures underlie the extent to which we meaningfully relate to nonhuman entities and seek to incorporate them into human institutions.

The penultimate section of this chapter brings us—at last—to rights. In particular, I review rights theories and explore how they relate to the concepts depicted in the above map. The objectives here are to understand the conceptual implications associated with alternate rights theories and to assess which theory provides a stronger basis for extending rights to nonhuman entities.

Hohfeld's lacuna: Will and interest theories of rights

While Hohfeld's framework specifies the kinds of legal relations that exist between entities, it does not explain the conditions under which entities might be *entitled* to rights. As mentioned above, Hohfeld held that legal relations were the province of humans alone.[26] He did not consider whether other types of entities might participate in such relations, thus denying the possibility that rights could be extended to nonhumans. This section seeks to address the gap in Hohfeld's work by introducing theories of rights that offer alternate accounts of their function and application with an eye towards assessing whether or not nonhuman entities might be eligible for rights.

Generally speaking, there are two main schools of thought on the purpose of rights. In will (or choice) theory, rights offer a vehicle for the demonstration of agency. Their purpose is to "afford the rights-holder the autonomy to control duties which are owed to her by others, thereby protecting individual autonomy" (Dodsworth et al., 2018, p. 3). Rights respect the choice a person makes "either negatively by not impeding it ... or affirmatively by giving legal or moral effect to it" (Finnis, 2011, p. 204). Only individuals who can elect to impose or waive duties that others have towards them are capable of holding rights (Marx & Tiefensee, 2015, p. 72). Clearly then, a certain level of cognitive functioning is required to execute such decisions, which limits the class of entities that may claim rights under this theory (Kurki, 2017, p. 79). Intuitively, this logic appears reasonable and even defensible. Rights demanded and exercised by the rights-bearer enjoy legitimacy. One theorist goes as far as to conclude that "[r]ights are *only* secure and effective when they are an expression of autonomy, the creation and possession of their bearers" (Ingram, 2008, p. 414; emphasis added). Therefore, will theory would seem to require that an entity possess certain ontological properties associated with the capacity for agency (i.e., consciousness, intelligence, intentionality, sentience, etc.) in order to qualify as a potential rights-holder.

Without referencing it directly, Shelton (2014) invokes the spirit of will theory when she elaborates on the characteristics of an entity that entitle it to rights:

> [r]ights ... presuppose autonomous and aware agents, capable of rational choice and moral deliberation, and thus capable of being held responsible for their actions. Such agents must also engage in the basic act of mutual recognition of shared moral agency, and thereby accept the rights claims of others.
>
> (p. 6)

Here, *legal* rights depend on a demonstration of *moral* agency, although, as discussed earlier, these concepts are not necessarily connected because they speak to different kinds of statuses and personhoods. One thing that interpretations of will theory have in common is their reliance on a menu of ontological properties from which a number of items are selected. The property(ies) selected become the basis(es) against which eligibility for rights is judged. The problems with this approach are that (1) there is little agreement among experts as to which property or combination of properties is sufficient to justify the extension of rights, and (2) finding a way to operationalize and empirically demonstrate the possession of certain properties has thus far proven elusive.

One might argue, however, that will theory is not fully foreclosed to the possibility of rights for nonhuman entities. In order to make a compelling case for rights eligibility, such a being would have to exhibit some semblance of autonomy, consciousness, rationality, and so on that could be readily interpreted as such by human observers. Thus, one could apply Danaher's (2020) "ethical behaviourism" (see Chapter One) to the realm of rights in an effort to determine which nonhuman entities act in ways that are performatively equivalent to those entities already deemed to possess rights on the basis of the aforementioned properties. On its face, externally approximating the possession of ontological properties poses less of an obstacle for humanoid robots operating under strong AI than it does for animals or nature. Should such a robot advocate for rights of its own, it might present humanity with a hard case for the extension of rights since "rights arise from, and must be based on, the activity of their bearers" (Ingram, 2008, p. 413). So far, neither animals nor nature have made such political overtures.

The opposing perspective is represented by interest (or benefit) theory, which stipulates that the purpose of rights is to protect a person's core interests. Instead of affording rights to only those entities capable of claiming or rejecting entitlements given their possession of certain properties, interest theory finds "that any subject which possesses or is capable of possessing interests may bear rights, as long as the corresponding interest is sufficiently important to justify ascribing duties onto others" (Dodsworth et al., 2018, p. 3). Applying this theory to Hohfeld's framework, interests and duties would form a pair of correlative incidents. Of course, such interests would need to be adequately defined and determined to be "sufficiently important" in order to justify requiring others to uphold concomitant duties. For Caney (2006), such "highly valued interests" include "liberty of conscience, association, and expression" (p. 259).

How do we know which and whose interests are integral to determining rights, and whether some interests might take precedence over others? By adapting a theoretical classificatory scheme developed by Lee (1999, pp. 184–185), interests can be arrayed along two sets of axes that identify a range of core concerns according to the centrality of humans and the scope of subjects involved—(1) anthropocentric versus non-anthropocentric, and (2) individualist versus holist. When considered in tandem, the positions along these axes can be combined, producing four distinct categories. I describe each position in terms of the interests involved and the extent to which it might permit rights for nonhumans.

To begin, anthropocentric individualism is an unlikely platform for pushing rights into the realm of the nonhuman. This position, emerging from the canon of Western philosophy in general and the work of Descartes in particular, prioritizes the individual human on the basis of interests intrinsic to the human species, such as autonomy and liberty. One interpretation of this position contends that *human* rights are grounded in possession of "normative agency," or "the capacity to choose and to pursue our conception of a worthwhile life" (Griffin, 2008, p. 45), and only humans (presently) have this ability. This capacity presupposes the possession of certain properties like consciousness and rationality. However, if a nonhuman entity could define its own conception of a worthwhile life (i.e., interests) and takes steps to pursue this vision, it might qualify for moral personhood, moral agency, and thus moral rights. Still, this approach seems unable to look beyond what are perceived as exclusively human interests that serve the ultimate purpose of safeguarding the uniquely human quality of normative agency.

Anthropocentric holism offers an arguably more encouraging route. One example of this position creates space for considering the interests of nonhumans, but only insofar as they relate to human values. This "weak anthropocentrism" (Norton, 1984) stipulates that nonhuman entities do not necessarily hold interests of their own that warrant protecting, but rather it is through the pursuit of human-centered interests (i.e., continued access to natural resources) that other beings such as ecosystems find themselves the objects of concern. Here, the focus is not so much on the qualities of nonhuman entities that suggest humans have moral obligations towards them. Instead, it is through the satisfaction of human interests that other kinds of entities might enjoy spillover effects that positively impact interests relevant to their flourishing or survival. This kind of indirect moral concern does not translate into rights for nonhumans, however, as such beings are not capable of fulfilling the minimum conditions necessary to possess rights (Norton, 1982).

Non-anthropocentric individualism identifies interests that apply across species but to singular entities. Interests related to this position include, *inter alia*, enjoying a quality life (Regan, 1987) and not being harmed (Pietrzykowski, 2017). Importantly, these interests stem from consciousness and sentience—ontological properties associated with psychological personhood. In terms of distinguishing rights theories, this is problematic, as the properties required for the possession of interests are some of the same found in will theory. If we accept that these properties are in some sense dispositive of an entity's eligibility for moral or legal consideration, then the rights theories become epiphenomenal to a certain set of characteristics, which would take on a more fundamental role in the determination of rights. At the risk of seeming redundant, I maintain that this position suffers from the same flaws that tarnish the utility of will theory—there is no agreement regarding which properties are the most important to have, and we lack ways of empirically verifying their presence in other beings.[27] For the moment, suffice it to say that non-anthropocentric individualism might only duplicate some of the challenges regarding the ascription of rights for nonhuman entities observed under will theory.

Finally, non-anthropocentric holism concerns the interests of all things comprising the whole, not just humans. Such interests include protecting the integrity, stability, and beauty of ecosystems (Naess, 1973; Fox, 1984), and promoting the wellbeing of all living beings (Taylor, 1981).[28] This position seems the most inclusive of the four, as potentially anything that advances interests pertaining to the maintenance of entire ecosystems could be entitled to rights designed to protect that function. Here, it is not the properties possessed by individual subjects that secure their admission to the community of rights bearers. Rather, rights serve the purpose of safeguarding the functioning of larger systems consisting of a wide range of entities, all of which play some part in preserving the whole. Although animals and nature are among the most obvious nonhuman members of this community, in principle anything that contributes to the vitality of a social-ecological system might possess rights. Moving from a strictly natural science-oriented ecology to a "radically relational ecology" (Coeckelbergh, 2010, p. 216) could expand the list of potentially rights-bearing entities to include nonhuman, inorganic technological beings.[29]

While Hohfeld is rightly credited with having defined the range of legal relations that may occur between subjects, his framework neglected to describe the kinds of entities eligible to partake in such relations. Reviewing the two major rights theories—will theory and interest theory—offers both clarity and complexity on the issue. Will theory unapologetically affords rights to only those capable of exercising their individual agency, which renders this approach vulnerable to the criticisms that we still don't know what property or mix of properties provides the soundest basis for rights, and that we don't have a universally accepted way of assessing the presence or absence of these properties. In addition, realizing that setting such a high bar might infringe upon the would-be rights of humans with cognitive deficiencies, courts have determined that such individuals still qualify for legal personhood and thus legal rights due to the inherent dignity they possess as humans (Cupp, 2017, p. 488). As I have argued, this circular reasoning does little to help us understand whether entities other than humans might enjoy rights.[30]

Interest theory suggests a wider array of possibilities. Positions relating to individualism entail interests that appear to require the possession of ontological properties, making them akin to will theory on a fundamental level. These positions leave little leeway for nonhuman entities aside from those who possess human-like cognitive capabilities that remain empirically challenging to define and verify. By contrast, positions promoting holism speak more broadly to, or are at least more ambivalent about, the kinds of interests integral to rights. These positions seem more amenable to extending rights to nonhumans.[31] While non-anthropocentric views are hospitable to the idea of interests possessed by nonhumans, even some anthropocentric perspectives find value in interest theory. For instance, one scholar holds that, of the two rights theories reviewed here, interests offer the "better view ... because otherwise humans who are incapable of making choices (such as infants, the handicapped and comatose) would not have rights" (Peters, 2016, p. 43). Overall, interest theory appears more likely than will theory to support the extension of rights to nonhuman entities.

Conclusion

The main objectives of this chapter were to clear up terminological inconsistencies among cognate concepts and demonstrate how they logically relate to one another. A subsidiary objective involved interrogating these concepts to assess the extent to which they might accommodate nonhuman entities. I have shown how different ontological properties or relational mechanisms are associated with different forms of personhood. Importantly, I have also argued that properties are themselves relational in nature, as are the conventional types of personhood. This argument has three implications for the assignment of rights. First, the traditional view in philosophy that what an entity *is* precedes how we *ought* to treat it is not as cleanly linear as many allege. Human tendencies and normative approaches to embracing alterity are intrinsically bound up with properties that entities reveal to us through the process of interaction. Thus far, no one has been able to definitively state which property(ies) is/are required for personhood and, eventually, rights. For the time being, it is enough to conclude that human relations with nonhuman entities are complex, contingent, and culturally determined, so to err on the side of caution, we should remain open to the possibilities that inhere in relationships of all kinds when questions of rights for nonhumans arise. Second, relational personhood undergirds both moral and legal status and, by extension, the designation of moral or legal rights. Whether an entity appears similar to us or useful for advancing our interests, it is the relational nature of the dyadic arrangement that determines whether the entity is worthy of rights. Finally, while extant theories of rights offer alternative views on the purpose of rights and the kinds of beings eligible for them, both will and interest theories are grounded in relations between those who determine the rights of others and those who have rights bestowed upon them. However, I contend that given the methodological issues associated with the verification of ontological properties, will theory and individualist positions within interest theory are more analytically troublesome for determining which entities qualify for rights. A more helpful approach can be found in holist positions on interest theory, although they vary in terms of the degree to which interests are based primarily on human concerns. A non-anthropocentric holist position potentially offers the strongest basis for extending rights to nonhuman and even non-living beings. The next task is to analyze the extent to which theory and practice regarding the rights of nonhuman living entities might inform the debate over rights for nonhuman technological ones.

Notes

1 French (1979) also identifies three forms of personhood, but he refers to what some call psychological personhood as "metaphysical" personhood (p. 207).
2 Although Scott (1990) is clearly describing moral persons, he uses this term synonymously with moral personhood (p. 81).
3 There are several different interpretations of intentionality. In its philosophical understanding, intentionality is defined as "that feature of certain mental states by which they are directed at or about objects and states of affairs in the world" (Searle, 2008, p. 333). In law, intentionality "is concerned more with the concept that people act for

reasons which they themselves control" (Calverley, 2008, p. 528). Folk psychology offers yet another definition, this time related to actions undertaken to achieve a desired outcome, assumptions about the consequences of those actions, and possession of the skills required to execute these actions.

4 But see Grear (2015), who argues that in fact "[t]he so-called 'natural person' of law is not an embodied, corporeally 'thick', flesh and blood human being at all, but a highly selective construct" (p. 237).

5 *Citizens United v. Federal Election Commission*, 130 S. Ct. 876 (2010).

6 *Santa Clara County v. Southern Pacific Railroad*, 118 U.S. 394 (1886).

7 Ibid., at 396.

8 It is worth mentioning that there exists a rift between the United States and other common law countries in terms of their legal basis for *in rem* proceedings. While the U.S. has long adhered to a personification doctrine, the U.K. and other common law jurisdictions have moved towards a procedural theory in which "the statutory right of action *in rem* is regarded as a procedural device to flush out the liable shipowner" (Myburgh, 2005, p. 283).

9 See *United States v. The Schooner Little Charles*, 26 F. Cas. 979, 982 (C.C.D. Va. 1818); *The Phebe*, 19 Fed. Cas. 424, No. 11,064 (D. Me. 1837).

10 *United States v. Brig Malek Adhel*, 43 U.S. (2 How.) 210 (1844).

11 Ibid., at 234.

12 Confusingly, the author refers to legal personhood as "rights of constitutional personhood" (Solum, 1992, p. 1255). I use the phrase *legal personhood* when discussing the author's argument for the sake of consistency.

13 For an extensive discussion of the differences between personhood, person, self, and identity in the context of anthropological research, see Appell-Warren (2014, Part Two).

14 In most cases, the philosophers discussed here do not actually employ the phrase *relational personhood*, as they are mainly referring to the moral status of nonhuman entities. However, what they are arguing for is a kind of moral consideration emerging from the relations between humans and other beings. Therefore, I am categorizing their writings on this subject, which find significant resonance with the anthropological works described earlier, as constituting arguments in favor of personhood established on a relational basis.

15 This does not mean, however, that Japanese society is more accepting of robots than other cultures or that the reasons for and impacts of such widespread adoption of robots are benign. One study showed that Japanese and American faculty members exhibit similar attitudes towards robots (MacDorman et al., 2009), while another found that Japanese respondents hold more negative attitudes towards robots than do Americans (Bartneck et al., 2005). In addition, given Japan's rapidly aging and shrinking population and antipathy towards foreign workers (Tomiura et al., 2019), some have surmised that introducing robots into Japanese society will help keep the country's economy afloat while avoiding the need to import migrant labor (Robertson, 2014). A skeptical view alleges that the notion of a uniquely Japanese receptivity to robots is deliberately advanced by academia, industry, and government, and researchers in particular feed into this trope through work that fails to reflexively examine its own underlying assumptions about cultural values and the desirability of robots in society (Šabanović, 2014, p. 360).

16 To wit, there is a famous robot hobbyist retail store in Akihabara, Tokyo, called Tsukumo Robot Kingdom. See https://robot.tsukumo.co.jp/.

17 But see Stone (1972), who dismisses Shinto myths as "quaint, primitive and archaic" (p. 498).

18 Deep ecology and transpersonal ecology are reviewed in greater detail in Chapter Four.

19 To these two criteria Wetlesen (1999) adds a third—"linguistic competence," which he argues "is necessary in order to understand the moral questions that are debated and the

answers given" (p. 298). However, for reasons related to a later argument described in Chapter Three, I leave this tertiary qualification aside for the moment.

20 As an example of the terminological inconsistency noted at the outset of this chapter, Arp (2005) uses the terms "person" and "personhood" interchangeably, despite the conceptual differences identified by anthropologists.

21 In order to avoid further confusion, I refrain from utilizing this latter category of legal entity in favor of adopting Turner's broader definition of a legal subject.

22 Theorists disagree about the extent to which Hohfeld's framework describes different kinds of legal relations or different manifestations of rights. While some (i.e., Wenar, 2005; Rainbolt, 2006; Gunkel, 2018; Turner, 2019) treat all Hohfeldian incidents as kinds of rights, others (i.e., Wellman, 1985) argue that only the entire suite of Hohfeldian incidents can constitute a right. For present purposes and following Hohfeld's original intentions, I do not treat all legal relations as synonymous with rights.

23 Hohfeld takes an agnostic position on the subject of competing privileges. This is a weakness of his admittedly descriptive framework.

24 The correlatives listed here represent the interpretive work of this author. Wellman did not explicitly enumerate pairs of moral incidents in his work.

25 Following Hohfeld's own synonyms, Wellman (1985) prefers to use the term "claim" instead of "right" and swaps "privilege" for "liberty."

26 But see Rainbolt (2006), who argues that a neo-Hohfeldian understanding of rights based on justified-constraint theory would not necessarily prohibit nonhuman entities (i.e., rocks) from having rights (p. 197).

27 I expand on this critique in Chapter Three, which focuses on animal rights.

28 Taylor (1981) explicitly notes that the interests he describes support the extension of legal, not moral, rights (p. 218).

29 I engage with this argument more fully in Chapter Four.

30 To wit, Wise (2013) maintains that certain animals possess dignity because of their capacity for practical autonomy. I expand on Wise's argument in Chapter Three.

31 Non-anthropocentric individualism and non-anthropocentric holism are explored in greater detail, respectively, in the context of animal rights (Chapter Three) and the rights of nature (Chapter Four).

References

Aaltola, E. (2008). Personhood and Animals. *Environmental Ethics, 30*(2), 175–193.

Andrade, F., Novais, P., Machado, J., & Neves, J. (2007). Contracting Agents: Legal Personality and Representation. *Artificial Intelligence and Law, 15*(4), 357–373.

Appell-Warren, L. P. (2014). *"Personhood": An Examination of the History and Use of an Anthropological Concept.* Edwin Mellen Press.

Arp, R. (2005). "If Droids Could Think …": Droids as Slaves and Persons. In K. S. Decker & J. T. Eberl (Eds.), *Star Wars and Philosophy: More Powerful Than You Can Possibly Imagine* (pp. 120–131). Open Court.

Bartneck, C., Nomura, T., Kanda, T., Suzuki, T., & Kato, K. (2005). Cultural Differences in Attitudes Towards Robots. *Proceedings of the AISB Symposium on Robot Companions: Hard Problems and Open Challenges in Human–Robot Interaction* (pp. 1–4).

Bird-David, N. (1999). "Animism" Revisited: Personhood, Environment, and Relational Epistemology. *Current Anthropology, 40*(S1), S67–S91.

Blackstone, W. (1766). *Commentaries on the Laws of England, Volume II, Of the Rights of Things.* Clarendon Press.

Blaser, M. (2014). Ontology and Indigeneity: On the Political Ontology of Heterogeneous Assemblages. *Cultural Geographies, 21*(1), 49–58.

Bryson, J. J. (2018). Patiency Is Not a Virtue: The Design of Intelligent Systems and Systems of Ethics. *Ethics and Information Technology*, *20*(1), 15–26.

Calverley, D. J. (2008). Imagining a Non-Biological Machine as a Legal Person. *AI and Society*, *22*(4), 523–537.

Caney, S. (2006). Cosmopolitan Justice, Rights and Global Climate Change. *Canadian Journal of Law and Jurisprudence*, *19*(2), 255–278.

Coeckelbergh, M. (2010). Robot Rights? Towards a Social-Relational Justification of Moral Consideration. *Ethics and Information Technology*, *12*(3), 209–221.

Coeckelbergh, M. (2011). Humans, Animals, and Robots: A Phenomenological Approach to Human–Robot Relations. *International Journal of Social Robotics*, *3*(2), 197–204.

Coeckelbergh, M. (2014). The Moral Standing of Machines: Towards a Relational and Non-Cartesian Moral Hermeneutics. *Philosophy and Technology*, *27*(1), 61–77.

Coeckelbergh, M., & Gunkel, D. J. (2014). Facing Animals: A Relational, Other-Oriented Approach to Moral Standing. *Journal of Agricultural and Environmental Ethics*, *27*(5), 715–733.

Cullinan, C. (2003). *Wild Law: A Manifesto for Earth Justice*. Green Books.

Cupp, R. L. (2017). Cognitively Impaired Humans, Intelligent Animals, and Legal Personhood. *Florida Law Review*, *69*(2), 465–518.

Danaher, J. (2020). Welcoming Robots into the Moral Circle: A Defence of Ethical Behaviourism. *Science and Engineering Ethics*, *26*, 2023–2049.

Darling, K. (2016). Extending Legal Protection to Social Robots: The Effects of Anthropomorphism, Empathy, and Violent Behavior Towards Robotic Objects. In R. Calo, A. M. Froomkin, & I. Kerr (Eds.), *Robot Law* (pp. 213–232). Edward Elgar.

Davies, M. (2007). *Property: Meanings, Histories, Theories*. Routledge-Cavendish.

Dennett, D. C. (1976). Conditions of Personhood. In A. O. Rorty (Ed.), *The Identities of Persons* (pp. 175–196). University of California Press.

Dewey, J. (1926). The Historic Background of Corporate Legal Personality. *Yale Law Journal*, *35*(6), 655–673.

Dodsworth, A., O'Doherty, S., & Oksanen, M. (2018). Introduction: Environmental Human Rights and Political Theory. In M. Oksanen, A. Dodsworth, & S. O'Doherty (Eds.), *Environmental Human Rights: A Political Theory Perspective* (pp. 1–16). Routledge.

Donnelly, J., & Whelan, D. J. (2018). *International Human Rights* (5th ed.). Westview Press.

Finnis, J. (2011). *Natural Law & Natural Rights* (2nd ed.). Oxford University Press.

Fowler, C. (2018). Relational Personhood Revisited. *Cambridge Archaeological Journal*, *26*(3), 397–412.

Fox, W. (1984). Deep Ecology: A New Philosophy of Our Time? *The Ecologist*, *14*(5–6), 194–200.

Fox, W. (1990). *Toward a Transpersonal Ecology: Developing New Foundations for Environmentalism*. Shambhala.

Fraser Jr., G. B. (1948). Actions In Rem. *Cornell Law Review*, *34*(1), 29–49.

French, P. A. (1979). The Corporation as a Moral Person. *American Philosophical Quarterly*, *16*(3), 207–215.

Gray, J. C. (1909). *The Nature and Sources of the Law*. Columbia University Press.

Gray, K., & Wegner, D. M. (2009). Moral Typecasting: Divergent Perceptions of Moral Agents and Moral Patients. *Journal of Personality and Social Psychology*, *96*(3), 505–520.

Grear, A. (2015). Deconstructing Anthropos: A Critical Legal Reflection on 'Anthropocentric' Law and Anthropocene 'Humanity.' *Law and Critique*, *26*(3), 225–249.

Griffin, J. (2008). *On Human Rights*. Oxford University Press.

Gunkel, D. J. (2012). *The Machine Question: Critical Perspectives on AI, Robots, and Ethics*. MIT Press.

Gunkel, D. J. (2018). *Robot Rights*. MIT Press.

Herbert, J. (1967). *Shintô: At the Fountain-Head of Japan*. Stein and Day.

Himma, K. E. (2009). Artificial Agency, Consciousness, and the Criteria for Moral Agency: What Properties Must an Artificial Agent Have to Be a Moral Agent? *Ethics and Information Technology*, *11*(1), 19–29.

Hohfeld, W. N. (1913). Some Fundamental Conceptions as Applied in Judicial Reasoning. *Yale Law Journal*, *23*(1), 16–59.

Hohfeld, W. N. (1917). Fundamental Legal Conceptions as Applied in Judicial Reasoning. *Yale Law Journal*, *26*(8), 710–770.

Holmes, Jr., O. W. (1881). *The Common Law*. Little, Brown, and Company.

Hubbard, F. P. (2011). "Do Androids Dream?": Personhood and Intelligent Artifacts. *Temple Law Review*, *83*, 405–474.

Ingram, J. D. (2008). What Is a "Right to Have Rights"? Three Images of the Politics of Human Rights. *American Political Science Review*, *102*(4), 401–416.

Jackson, S. E. (2019). Facing Objects: An Investigation of Non-Human Personhood in Classic Maya Contexts. *Ancient Mesoamerica*, *30*(1), 31–44.

Johnson, D. G. (2006). Computer Systems: Moral Entities but Not Moral Agents. *Ethics and Information Technology*, *8*(4), 195–204.

Jones, R. A. (2013). Relationalism through Social Robotics. *Journal for the Theory of Social Behaviour*, *43*(4), 405–424.

Jones, R. A. (2016). *Personhood and Social Robotics: A Psychological Consideration*. Routledge.

Kelch, T. G. (1999). The Role of the Rational and the Emotive in a Theory of Animal Rights. *Boston College Environmental Affairs Law Review*, *27*(1), 1–41.

Kens, P. (2015). Nothing to Do With Personhood: Corporate Constitutional Rights and the Principle of Confiscation. *Quinnipiac Law Review*, *34*(1), 1–37.

Kitano, N. (2006). "Rinri": An Incitement Towards the Existence of Robots in Japanese Society. *International Review of Information Ethics*, *6*, 78–83.

Knauer, N. J. (2003). Defining Capacity: Balancing the Competing Interests of Autonomy and Need. *Temple Political and Civil Rights Law Review*, *12*, 321–347.

Koops, B.-J., Hildebrandt, M., & Jaquet-Chiffelle, D.-O. (2010). Bridging the Accountability Gap: Rights for New Entities in the Information Society? *Minnesota Journal of Law, Science and Technology*, *11*(2), 497–561.

Kurki, V. A. J. (2017). Why Things Can Hold Rights: Reconceptualizing the Legal Person. In V. A. J. Kurki & T. Pietrzykowski (Eds.), *Legal Personhood: Animals, Artificial Intelligence and the Unborn* (pp. 69–89). Springer.

Kurki, V. A. J., & Pietrzykowski, T. (2017). Introduction. In V. A. J. Kurki & T. Pietrzykowski (Eds.). *Legal Personhood: Animals, Artificial Intelligence and the Unborn* (pp. vii–ix). Springer.

Lee, K. (1999). *The Natural and the Artefactual: The Implications of Deep Science and Deep Technology for Environmental Philosophy*. Lexington Books.

Lehman-Wilzig, S. (1981). Frankenstein Unbound. *Futures*, *13*(6), 442–457.

Lewis, J. E., Arista, N., Pechawis, A., & Kite, S. (2018). Making Kin with the Machines. *Journal of Design and Science*, *3*, 5. doi:10.21428/bfafd97b.

Lind, D. (2009). Pragmatism and Anthropomorphism: Reconceiving the Doctrine of the Personality of the Ship. *University of San Francisco Maritime Law Journal*, *22*(1), 39–122.

Locke, J. (1836). *An Essay Concerning Human Understanding* (27th ed.). T. Tegg and Son.

MacDorman, K. F., Vasudevan, S. K., & Ho, C.-C. (2009). Does Japan Really Have Robot Mania? Comparing Attitudes by Implicit and Explicit Measures. *AI and Society*, *23*(4), 485–510.

Manus, P. (1998). One Hundred Years of Green: A Legal Perspective on Three Twentieth Century Nature Philosophers. *University of Pittsburgh Law Review*, *59*(3), 557–676.

Manzotti, R., & Jeschke, S. (2016). A Causal Foundation for Consciousness in Biological and Artificial Agents. *Cognitive Systems Research*, *40*(C), 172–185.

Marx, J., & Tiefensee, C. (2015). Of Animals, Robots and Men. *Historical Social Research/ Historische Sozialforschung*, *40*(4), 70–91.

Matambanadzo, S. M. (2013). The Body, Incorporated. *Tulane Law Review*, *87*(3), 457–510.

Mawani, R. (2018). Archival Legal History: Towards the Ocean as Archive. In M. D. Dubber & C. Tomlins (Eds.), *The Oxford Handbook of Legal History* (pp. 291–310). Oxford University Press.

Mitsukuni, Y., Ikko, T., & Tsune, S. (Eds.) (1985). *The Culture of ANIMA*. Mazda Motor Corporation.

Myburgh, P. (2005). Arresting the Right Ship: Procedural Theory, the In Personam Link and Conflict of Laws. In Martin Davies (Ed.), *Jurisdiction and Forum Selection in International Maritime Law: Essays in Honor of Robert Force* (pp. 283–320). Kluwer Law International.

Naess, A. (1973). The Shallow and the Deep, Long-Range Ecology Movement: A Summary. *Inquiry*, *16*(1–4), 95–100.

Naess, A. (1995). Self-Realization: An Ecological Approach to Being in the World. In G. Sessions (Ed.), *Deep Ecology for the 21st Century: Readings on the Philosophy and Practice of the New Environmentalism* (pp. 225–239). Shambhala.

Neuman, S. (2018, May 1). Japan, Old Robot Dogs Get A Buddhist Send-Off. *NPR*. https://www.npr.org/sections/thetwo-way/2018/05/01/607295346/in-japan-old-robot-dogs-get-a-buddhist-send-off.

Norton, B. G. (1982). Environmental Ethics and Nonhuman Rights. *Environmental Ethics*, *4*(1), 17–36.

Norton, B. G. (1984). Environmental Ethics and Weak Anthropocentrism. *Environmental Ethics*, *6*(2), 131–148.

Peters, A. (2016). Liberté, Égalité, Animalité: Human–Animal Comparisons in Law. *Transnational Environmental Law*, *5*(1), 25–53.

Pietrzykowski, T. (2017). The Idea of Non-Personal Subjects of Law. In V. A. J. Kurki & T. Pietrzykowski (Eds.), *Legal Personhood: Animals, Artificial Intelligence and the Unborn* (pp. 49–67). Springer.

Pollock, J. L. (1989). *How to Build a Person: A Prolegomenon*. MIT Press.

Rainbolt, G. W. (2006). *The Concept of Rights*. Springer.

Regan, T. (1987). The Case for Animal Rights. In M. W. Fox & L. D. Mickley (Eds.), *Advances in Animal Welfare Science 1986/87* (pp. 179–189). Martinus Nijhoff Publishers.

Robertson, J. (2014). Human Rights vs. Robot Rights: Forecasts from Japan. *Critical Asian Studies*, *46*(4), 571–598.

Šabanović, S. (2014). Inventing Japan's "Robotics Culture": The Repeated Assembly of Science, Technology, and Culture in Social Robotics. *Social Studies of Science*, *44*(3), 342–367.

Sahlins, M. (2011). What Kinship Is (Part One). *Journal of the Royal Anthropological Institute, 17*(1), 2–19.

Schodt, F. L. (1988). *Inside the Robot Kingdom: Japan, Mechatronics, and the Coming Robotopia.* Kodansha International.

Scott, G. E. (1990). *Moral Personhood: An Essay in the Philosophy of Moral Psychology.* State University of New York Press.

Searle, J. R. (2008). Minds, Brains, and Programs. In J. Feinberg & R. Shafer-Landau (Eds.), *Reason and Responsibility: Readings in Some Basic Problems of Philosophy* (13th ed., pp. 330–342). Thomson Wadsworth.

Shelton, D. L. (2014). *Advanced Introduction to International Human Rights Law.* Edward Elgar.

Shestack, J. J. (1998). The Philosophic Foundations of Human Rights. *Human Rights Quarterly, 20*(2), 201–234.

Smith, B. (1928). Legal Personality. *Yale Law Journal, 37*(3), 283–299.

Solaiman, S. M. (2017). Legal Personality of Robots, Corporations, Idols and Chimpanzees: A Quest for Legitimacy. *Artificial Intelligence and Law, 25*(2), 155–179.

Solum, L. B. (1992). Legal Personhood for Artificial Intelligences. *North Carolina Law Review, 70*(4), 1231–1288.

Søraker, J. H. (2007). The Moral Status of Information and Information Technology: A Relational Theory of Moral Status. In S. Hongladarom & C. Ess (Eds.), *Information Technology Ethics: Cultural Perspectives* (pp. 1–19). Idea Group.

Splitter, L. J. (2015). *Identity and Personhood: Confusions and Clarifications Across Disciplines.* Springer.

Stone, C. D. (1972). Should Trees Have Standing? Toward Legal Rights for Natural Objects. *Southern California Law Review, 45*, 450–501.

Sullins, J. P. (2006). When Is a Robot a Moral Agent? *International Review of Information Ethics, 6*, 23–30.

Sumner, L. W. (1987). *The Moral Foundation of Rights.* Clarendon Press.

Tasioulas, J. (2019). First Steps Towards an Ethics of Robots and Artificial Intelligence. *Journal of Practical Ethics, 7*(1), 61–95.

Taylor, P. W. (1981). The Ethics of Respect for Nature. *Environmental Ethics, 3*(3), 197–218.

Teubner, G. (2006). Rights of Non-Humans? Electronic Agents and Animals as New Actors in Politics and Law. *Journal of Law and in Society, 33*(4), 497–521.

Tomiura, E., Ito, B., Mukunoki, H., & Wakasugi, R. (2019). Individual Characteristics, Behavioral Biases, and Attitudes Toward Foreign Workers: Evidence from a Survey in Japan. *Japan and the World Economy, 50*, 1–13.

Turner, J. (2019). *Robot Rules: Regulating Artificial Intelligence.* Palgrave Macmillan.

Vallverdú, J. (2011). The Eastern Construction of the Artificial Mind. *Enrahonar: Quaderns de Filosofia, 47*, 171–185.

Vermeylen, S. (2017). Materiality and the Ontological Turn in the Anthropocene: Establishing a Dialogue Between Law, Anthropology and Eco-Philosophy. In L. J. Kotzé (Ed.), *Environmental Law and Governance for the Anthropocene* (pp. 137–162). Hart Publishing.

Vincent, A. (1989). Can Groups Be Persons? *The Review of Metaphysics, 42*(4), 687–715.

Watson, R. A. (1979). Self-Consciousness and the Rights of Nonhuman Animals and Nature. *Environmental Ethics, 1*(2), 99–129.

Wellman, C. (1985). *A Theory of Rights.* Rowman & Allanheld.

Welters, M. (2013). Towards a Singular Concept of Legal Personality. *Canadian Bar Review*, *92*, 417–455.

Wenar, L. (2005). The Nature of Rights. *Philosophy and Public Affairs*, *33*(3), 223–252.

Wetlesen, J. (1999). The Moral Status of Beings Who Are Not Persons: A Casuistic Argument. *Environmental Values*, *8*(3), 287–323.

Wise, S. M. (2010). Legal Personhood and the Nonhuman Rights Project. *Animal Law*, *17*(1), 1–12.

Wise, S. M. (2013). Nonhuman Rights to Personhood. *Pace Environmental Law Review*, *30*(3), 1278–1290.

Youatt, R. (2017). Personhood and the Rights of Nature: The New Subjects of Contemporary Earth Politics. *International Political Sociology*, *11*(1), 39–54.

3 The rights of animals

In search of humanity

> After all, how little we know of the inner life of animals. How few our facts are, and how little certain we are of them.
>
> (Archibald Banks, 1874, p. 809)[1]

Having clarified the content of concepts that lead to rights and how such concepts relate to each other, we now possess firm grounding for exploring theory and practice regarding rights for certain classes of nonhumans. This chapter focuses on animals. It seeks to understand the extent to which the debate over animal rights can inform responses to the machine question. In order to accomplish this task, I review religious beliefs and intellectual positions about the treatment of animals; compare properties-based, direct/indirect, relational, and legal approaches to animal rights; and examine recent cases in which petitioners attempt to secure rights for nonhuman creatures. I argue that while properties-based and direct/indirect arguments have resulted in an impasse unlikely to be overcome any time soon, relational approaches offer a more promising avenue for animal rights. Further, I explain how issues involving anthropocentrism, duties, properties, and inclusiveness commonly observed within scholarship and case law on animal rights frustrate the translation of this model to the realm of technology. Finally, I conclude by describing the conditions under which philosophy and law on animal rights might contribute to the discussion regarding rights for robots.

The treatment of animals:
Religious and intellectual perspectives

The treatment of animals has been addressed by a variety of faiths since antiquity and debated among intellectuals for centuries. Both Western and Eastern religions have advanced a variety of perspectives on the subject, and the ideas put forth during the Enlightenment continue to influence conversations about animal ethics today. In the space below, I offer a brief overview of how major religious traditions have characterized their respective ethical orientations towards animals and then summarize arguments about the status of animals according to Enlightenment thinkers whose writings have informed discussions on animal rights.

The view found among members of the Jewish faith stems from the Old Testament and rabbinic interpretations of the ancient tome, which advocate in favor of treating animals with compassion while also permitting the slaughter of animals under certain restricted conditions (Vogel, 2001). The Christian thought about animals, based largely on Biblical scripture and sectarian theology, is often framed in terms of the unique place humans occupy in the creation of the world and the dominion they enjoy over beasts of the land and sea (Linzey & Regan, 1990, p. 3). Although for Catholics animals are seen as having neither moral status nor moral rights (Linzey, 1989, p. 134), a more charitable reading of the Bible suggests that humans should treat animals with love and compassion due to our moral responsibility as creatures created in the image of God (Regan, 1990; Phelps, 2002). Despite the diversity observed among its sub-traditions, Buddhism in general affords ethical significance to animals other than humans (Waldau, 2000). In particular, the First Precept of Buddhism, "Do not Kill," applies to humans and animals in equal measure (Phelps, 2004, pp. 48–49). Jainism, while ordaining the existence of an ontological divide between humans and animals, nevertheless promotes the ideas that all sentient beings hold the potential for self-transcendence and that humans must treat nonhumans in a conscientious and nonviolent manner (Vallely, 2014, p. 52). Hindu faiths support the extension of moral standing to animals on the basis that they possess a good of their own, which can be harmed or advanced by humans (Framarin, 2014, p. 42). However, conflicting interpretations of ancient texts suggest that the Hindu perspective on human–animal relationships is not as uniformly benevolent as many allege (Mawdsley, 2006, p. 384). Buddhism, Jainism, and Hindu traditions alike all promote the doctrine of *ahiṃsā*, or "non-injury to all living things and reverence for all life" (DeGrazia, 2002, p. 6), which is often described as providing a religious foundation for the compassionate treatment of animals. In Confucian thought, humans retain a position of superiority over animals based on the notion that they are naturally closer to one another than they are to nonhumans. Although beings of all kinds hold moral value, the worth of animals is assessed according to "how they affect or are affected by human beings" (Bao-Er, 2014, p. 88). Followers of Islam observe that while Allah created the world for the benefit of humans, the resulting power they have over animals comes with certain responsibility. The Qu'ran stipulates that animals should only be killed when absolutely necessary, and the very presence of animals serves to compel "humans to reflect upon the divine Beneficence they receive" (Szűcs et al., 2012, p. 1503). In short, religions across the world convey a diversity of views regarding the relationship between humans and animals, although all arguably share a common capacity for doctrinally derived moral and ethical consideration for members of the animal kingdom. However, the degree to which such religions provide theological footing for a non-anthropocentric ethic that affords animals a moral status on par with that bestowed upon humans is still the subject of much debate.

Several notable figures central to the advancement of human reason during the Enlightenment have commented on the place of animals in human ethical systems. These ideas have persisted well beyond their birth, greatly influencing the tenor of contemporary discourses surrounding animal ethics and animal rights. I briefly recount some of the most significant contributions to these discourses made by European philosophers, which set the intellectual backdrop for modern ethical debates about animals. To begin, French mathematician René Descartes (1596–1650) proposed a view of animals that comports with a philosophical outlook defined by strict dichotomies—mind/body, man/nature, and nature/culture. According to Descartes, animals consist of a body without a mind, whereas humans possess both. Bereft of reason, language, or a soul, animals are more like machines than people (Descartes, 1637/1924, pp. 59–63). For instance, the seasonal migration of swallows demonstrates that they operate with the mechanical precision of a clock (Descartes, 1970, p. 207). This view has come to be known as Descartes' doctrine of the "animal machine" (*bête-machine*) (Cottingham, 1978, p. 551). However, his use of a mechanistic analogy should not be taken to mean that Descartes believed that animals lack the capacity to feel. On the contrary, his writings indicate that he remained "cautiously agnostic on the whole question" (P. Harrison, 1992, p. 227).

German philosopher Immanuel Kant (1724–1804) argued that animals were "not mere machines" because they have a soul that causes the matter comprising them to become animate (Kant, 1997, p. 86). However, this conclusion does not suggest that animals possess inherent worth or independent moral value, qualities that justify the receipt of ethical concern. For Kant, responsibility towards animals is intrinsically bound up with our interest in preserving human dignity (Wilson, 2017, p. 12). The reasoning proceeds as follows. The manner in which humans treat animals affects how we come to treat each other. When we practice kindness towards animals, we cultivate good moral behavior that guides our interactions with other humans. By learning to treat each other well, we advance human dignity. As Kant explained, "[t]ender feelings towards dumb animals develop humane feelings towards mankind" (Kant, 1963, p. 240). Importantly, this approach holds that animals are not valued for their own sake because they possess a soul. After all, they "are not self-conscious and are there merely as a means to an end" (Kant, 1963, p. 239). For Kant, humans do not have direct duties towards animals. Our treatment of animals reflects indirect duties we have towards other *humans*.

The final voice of Enlightenment thinking regarding animals discussed here is English philosopher Jeremy Bentham (1748–1832). While Bentham wrote little about animals and is more well known for founding the moral theory of utilitarianism, his brief remarks on the topic greatly influenced arguments made by animal liberation advocates such as Peter Singer.[2] In the span of a long footnote, Bentham raised four contentions about animals in the context of a comment on the kinds of entities capable of experiencing happiness. First, upon questioning why animals had not been afforded protection through legislation, he concluded that there is no good reason for this omission other than human fear and the inability of animals to convey their interests. Second, Bentham found that withholding rights on the basis that a being is an animal was just as unjustifiable as doing so because of a person's skin color, given that both animals and slaves are sensitive beings. Third, he pointed out that using rationality as a benchmark for ethical treatment is problematic, as some animals are more rational than human infants.[3] Fourth, Bentham proposed his own criterion for ethical consideration—a being's capacity to feel pain. His now-famous line of inquiry stated that "the question is not, Can they *reason*? nor, Can they *talk*? But, Can they *suffer*?" (Bentham, 1879, p. 311; emphasis in original). Although Bentham was by one account "the first Western philosopher to grant animals equal consideration from within a comprehensive, non-religious moral theory," he did not reject the practice of killing animals "as long as a pointless cruelty could be avoided" (Kniess, 2019, p. 556). Thus, while Bentham is widely credited with having contributed positively to the case for animal rights, this legacy glosses over the complexity of his perspective about animal treatment more generally.

The Enlightenment resulted in some significant developments in the area of animal ethics that were largely untethered from religious doctrine. While Descartes argued for a clean separation between humans and animals by virtue of the latter's mechanical, not mental, existence, Kant created space for the ethical consideration of animals, but it was carved out in unmistakably anthropocentric terms. Bentham's relatively minor contribution has had perhaps the most profound impact of the three perspectives recounted here. He posited that the capacity for suffering was the true measure by which an entity's eligibility for ethical treatment under the law should be assessed. As we shall see in the next section, these ideas have continued to shape conversations about animal ethics as they pertain to concepts such as personhood, status, and rights.

Animal rights: Properties-based, direct/ indirect, relational, and legal approaches

The field of animal ethics has long had as its central (normative) goal deriving a philosophical formula for the substantiation of animal rights. In some ways, the literature on this topic has led to important insights that have invigorated broader discussions about what it means to be human and the scope of the moral circle. In other ways, its focus on the qualities that animals share with humans has almost inevitably fostered a rather stale debate over the type of human trait that offers the most compelling grounds for treating animals like more than mere machines. In this section, I review arguments from philosophy and law that address the question of animal rights. I also discuss efforts that seek to apply lessons from the animal rights debate to the machine question. I close by revealing some of the terminological, conceptual, and empirical issues that aggravate the ability to obtain consensus and move forward with practical implementation of animal rights.

Most arguments in favor of extending rights to animals adopt a non-anthropocentric, individualist stance. That is, humans are not the only entities worthy of rights, which are bestowed upon individuals, not collectivities such as ecosystems.[4] However, the underlying reasoning employed in these arguments is often implicitly anthropocentric, as it hinges on how well animals approximate humanlike qualities (i.e., properties-based approaches); how the treatment of animals is guided by human duties or emotions (i.e., direct/indirect approaches); how the mechanisms by which animals are recognized involve human tendencies or action (i.e., relational approaches); or a range of human-centered ideas about the status of animals in the domain of law (i.e., legal approaches). As shown in the previous chapter, the determination of rights begins with an accounting of properties or mechanisms that leads to a given form of personhood, followed by the evaluation of an entity's moral or legal status. Figuring out which rights apply in accordance with a certain rights theory constitutes the final step along this path of inquiries. As we shall see, the literature on animal ethics makes an Olympic sport out of jumping around this sequentially organized conceptual scheme. In practical terms, this makes it very difficult to compare the various approaches to animal rights with any kind of logical consistency.

Properties-based approaches to animal ethics have largely dominated the debate over animal rights. One approach develops a positive case for the moral status of animals by arguing that the presence of certain attributes justifies the expansion of the moral circle to include nonhuman beings. Another approach advances a negative case that denies moral status to animals because of properties they are said to lack (Zamir, 2007, p. 18). Two of the most famous arguments in favor of granting rights to animals find home in the former camp. Singer (1974), divining inspiration from Bentham, maintains that sentience—understood here as the ability to experience suffering or happiness—is the trait possessed by both humans and animals that qualifies an entity for rights. Only sentient creatures, the argument goes, are capable of having interests that can be fulfilled or frustrated. It is on the basis of possessing cognizable interests that the extension of rights is legitimated.

Regan (1987) takes a broader view, locating in both animals and humans a common state of existence—being the subject-of-a-life. This capacity for real lived experience requires consciousness, sentience, and other abilities such as autonomy, belief, desire, intentionality, and memory (Regan, 1983, p. 153). Entities in possession of these properties hold inherent value, thus entitling them to rights. Animals that exhibit consciousness and sentience but which lack other mental capacities are still worthy of moral concern, but not as much as that which is afforded those in possession of the larger suite of abilities mentioned above. However, both of these groups—"higher" and "moderate" objects worthy of moral concern, respectively (Sun, 2018, p. 547)—are considered moral patients, as opposed to moral agents.

Regan also dismisses the contractarian argument that only beings rational enough to understand and accede to the terms of a social contract are worthy of moral consideration. Under this logic, the only basis for concern regarding non-rational entities is the sentimental value they hold for people. At best, this sentimentality suggests that humans have only indirect duties towards animals, because ultimately what matters are the feelings of other humans, not of animals themselves. As an exemplar of the positive case, Regan's (2004) argument rests on the observation that animals share with humans a number of morally relevant properties, including common language, behavior, bodies, systems, and origins (pp. 54–8).

Rowlands (1997) refutes the dismissal of the contractarian argument through an analysis of Rawlsian contractarianism. He concludes that rationality is only required of the architects of the contract, not its referents. As such, rationality is morally arbitrary and cannot function as the litmus test for the possession of moral personhood and thus moral rights. However, in agreement with Singer, Rowlands contends that sentience is the appropriate criterion for determining moral considerability. Since animals are sentient, even under a contractarian approach they would be eligible for moral rights.

The import of sentience is again evoked by Pietrzykowski (2017), who concurs with Regan that possession of this quality makes animals subjects-of-a-life. Sentience entails "subjective mental states" that alert animals to how their "existence may be better or worse for them" (Pietrzykowski, 2017, p. 58). Given their capacity to be aware of their present condition, animals may be said to have interests of their own, entitling them to legal rights. However, this does not mean that animals should be treated as nonhuman legal persons; instead, they should be deemed members of an entirely separate category—"non-personal subjects of law" (Pietrzykowski, 2017, p. 59). This maneuver inoculates against the problems associated with drawing a rhetorical analogy to humans and results in the extension of a single right, the "right to have one's own individual interests considered as relevant in all decisions that may affect their realization" (Pietrzykowski, 2017, p. 59). However, this category might only pertain to vertebrate animals whose sentience suggests that they have an inherent good and an interest in obtaining that good.

Other theorists have claimed that animals possess interests and therefore deserve rights, but the means by which they arrive at that conclusion varies. Feinberg (2013) lays out a philosophical argument that he develops in the context of individual animals and then applies to a number of borderline cases for rights eligibility, including, *inter alia*, vegetables, dead persons, and future generations. The logical pathway he derives from the animal case highlights the importance, in the first instance, of having a "conative life," which involves a being's possession of "conscious wishes, desires, and hopes ... urges and impulses ... unconscious drives, aims, and goals ... [and] latent tendencies, direction of growth, and natural fulfillments" (Feinberg, 2013, p. 374). Interests, which are crucial to determining whether or not an entity has a good of its own, arise out of such conations. Therefore, the capacity for conative life dictates whether an animal has interests worth protecting. The author takes one final step, which is to assert that if humans agree that these interests should be safeguarded for the animal's sake and not just to placate human sensibilities, it stands to reason that animals must also have rights.

Wise (2013), the famed animal rights litigator who heads the Nonhuman Rights Project (NhRP),[5] approaches the subject pragmatically, not philosophically,[6] through inductive reasoning. Observing the logic employed by common law judges in modern American cases, he surmises that autonomy, the factor underlying dignity, appears to be the main characteristic by which the potential to possess rights is evaluated.[7] More specifically, Wise suggests that the determination of rights hinges on demonstrating "practical autonomy," which comprises cognitive complexity, intentionality, consciousness, and sentience. However, positive demonstration of practical autonomy is only a sufficient, not necessary, condition for legal personhood. Entities that satisfy the criteria for practical autonomy may be said to possess fundamental interests and legal personhood, which entitles them to so-called "liberty rights" to "bodily integrity and bodily liberty" (Wise, 2002, p. 38).

But not all animals are created equal. Acknowledging the diversity present among members of the animal kingdom, Wise devises a continuous measurement of autonomy that ranges from 0.0 (least autonomous) to 1.0 (most autonomous). In accordance with their autonomy value, animals are then grouped into four discrete categories that indicate progressively greater legal consideration. Animals with a score ranging from 0.0 to 0.49 (Category 1) have little autonomy and thus do not possess liberty rights, whereas animals with a score ranging from 0.9 to 1.0 (Category 4) exhibit a level of autonomy that is sufficient for the enjoyment of such rights. This non-dichotomous classification scheme pushes the discussion of animal rights in a direction not often found in the pages of philosophical texts by recognizing the complexity and variation found in the nonhuman world as opposed to applying bright either/or lines.

To be sure, not all philosophers respond to the animal question by advancing a positive case for their moral status. There are those who espouse the negative case, in which the absence of some morally relevant property prohibits animals from joining the moral circle.[8] One of the more prominent torchbearers of this approach is Frey (1980), who forcefully argues against the idea that animals have interests that justify granting them moral rights. He rejects this premise for three reasons. First, *contra* Feinberg, Frey contends that while animals indeed possess interests in terms of having a good or wellbeing that can be harmed or helped by the actions of another, this assertion is logically problematic as a pathway for establishing their rights because it would mean that human artefacts might also have interests and thus moral rights. Therefore, due to the slippery slope that would seem to accept that even inanimate objects have interests, this line of reasoning is fatally flawed. Second, if interests stem from the possession of wants that can be fulfilled or denied, animals do not hold such interests because they have physical needs in the form of biological imperatives that serve to promote survival, not conative wants. If needs were synonymous with wants, everything from tractors to trees to tarantulas would have interests and a claim to moral rights. Again, the slippery slope renders animal interests logically untenable. Third, *contra* Regan, Frey finds that the absence of an intelligible language exhibited by creatures in the animal kingdom casts doubt on their case for having interests. His logic proceeds thusly: without language, you cannot have beliefs; without beliefs, you cannot have desires; without desires, you cannot have interests; and without interests, you are not entitled to moral rights. Further, behavior alone does not provide a sufficient basis for determining whether or not an animal grasps beliefs (Frey, 1980, p. 114).

Interrogation of the negative case shines a light on the role that human exceptionalism plays in the determination of rights where nonhuman entities are concerned.

If humans deserve rights even when they lack some property alleged to distinguish them from brutes (i.e., autonomy, consciousness, intelligence, intentionality, rationality, sentience, etc.), why can't beings in possession of such traits qualify for moral consideration? This line of inquiry is known as the "argument from marginal cases" (Narveson, 1977, p. 164). The answer to this question forces us to draw the moral circle with clearer boundaries by "demand[ing] consistency in our thinking about animals" (Tanner, 2009, p. 52). Both Singer and Regan propose versions of the argument from marginal cases that they deploy in order to strengthen their logical foundations for the extension of rights to animals. Singer (1979) holds that all sentient beings, be they animals or intellectually disabled humans, have interests under a basic principle of equality in which "we give equal weight in our moral deliberations to the like interests of all those affected by our actions" (p. 19). This "[e]qual consideration of interests allows us to treat different beings differently, but only when their interests differ" (Dombrowski, 1997, p. 20). Living entities (i.e., all sentient beings) are qualitatively different from persons, who are rational, self-conscious beings that possess self-awareness (Singer, 1979, p. 78). Marginal cases such as impaired humans would qualify as sentient beings but not persons, whereas great apes would be both sentient beings and persons. The practical effect of this distinction is that mentally undeveloped humans could be subject to the same treatment that would befall lower animals because their interests would be similar and neither would be considered persons. While this conclusion would no doubt seem noxious to the sensibilities of most people, Singer offers the rejoinder that it is our attitudes towards intellectually disabled human beings that need to change. To briefly summarize, Singer's view on the argument from marginal cases rests on three claims: (1) sentience serves as the floor for moral consideration; (2) equality dictates that all living beings should be treated the same on the basis of their interests; and (3) persons have interests different from those of merely living entities. Therefore, the way to deal with such cases is to treat like beings equally, even if we aren't particularly enthusiastic about the practical implications of such an approach.

Regan tackles the argument from marginal cases from a slightly different position. As mentioned above, Regan (1983) promotes the idea that some animals are subjects-of-a-life, which qualifies them for moral status and therefore moral rights. His reasoning proceeds thusly:

P_1: Some animals have beliefs, desires, perception, memory, preferences, interests, etc.;

P_2: All beings that have beliefs, desires, perception, memory, preferences, interests, etc., are subjects-of-a-life;

P_3: All subjects-of-a-life have inherent value;

P_4: All beings with inherent value are moral agents or moral patients;

P_5: All moral agents or moral patients are entitled to moral rights;

C: Some animals are entitled to moral rights.

En route to establishing a philosophical basis for animal rights, Regan (1979) introduces two ways of conceiving arguments from marginal cases: "(1) certain animals have certain rights because these [marginal] humans have these rights" or "(2) if these [marginal] humans have certain rights, then certain animals have these rights also" (p. 189). The response to these arguments depends on the quality possessed even by marginal humans that entitles them to moral rights. If, in the absence of other traits, such humans still have moral rights, it demonstrates that those traits are not necessary for the enjoyment of said rights. Comparing candidates in the running for "most reasonable criterion for right-possession" (Regan, 1979, p. 189), the author anoints inherent value as the victor. Therefore, both some animals and marginal humans have moral rights because of their inherent value that is independent of the interests held by others. Of course, inherent value depends on the possession of qualities associated with being a subject-of-a-life.

Cupp (2017), writing from the perspective of law and not philosophy, disagrees with the conclusions reached by both Singer and Regan on practical grounds. His refutation focuses on how courts and legislatures have addressed the legal rights of cognitively impaired humans. Cupp contends that legal institutions have overwhelmingly relied on the concept of human dignity, not cognitive capabilities, when assessing the legal personhood and legal rights of marginal people. Furthermore, the consequences of using cognitive factors as the basis for animal rights "could unintentionally lead to gradual erosion of protections for these especially vulnerable humans" (Cupp, 2017, p. 499), whom courts might be tempted to view as mentally equivalent to animals. Interestingly, it is this very same dignity that Wise (2013) argues could provide a platform for animal rights because of its underlying emphasis on autonomy, which he asserts some animals (i.e., chimpanzees and elephants) possess.

Analyzing the argument from marginal cases is an important intellectual exercise. It helps us to understand which criteria are not essential to an entity's case for moral rights. Recognizing that some humans will lack certain properties that philosophers have associated with moral status, the question then becomes—at what point in the process of eliminating morally relevant traits do animals or other nonhumans possess enough similarity with people to warrant including them in the moral circle? The danger of developing a rubric that could potentially justify harm to humans figures prominently in the debate over the correct answer to this question. At the same time, animal rights theorists have found the exclusion of creatures from the moral circle equally offensive. The difficulty inherent in resolving this quandary stems from the employment of an epistemology that emphasizes the significance of properties.

As mentioned in Chapter Two, properties-based approaches to defining the moral status of nonhuman beings are inherently subjective and, thus far, suffer from a lack of consensus.[9] As Nagel (1974) discusses in his provocatively titled article "What Is It Like to Be a Bat?", it is nearly impossible to access another entity's "subjective character of experience" without relying on our own imagination (p. 436). However, a few scholars have sought to address critical flaws of properties-based approaches in the hopes of paving a productive way forward. Chan (2011) offers one such suggestion. He argues that there is a non-zero probability that a living organism possesses sentience and consciousness (i.e., these properties are continuous, not dichotomous), and that uncertainty about the extent to which these properties are present in a being is not a valid reason for excluding it from moral consideration. Instead of gatekeeping the moral circle on an all-or-nothing basis, humans should scale their moral responsibilities towards other organisms according to the likelihood that they hold morally relevant traits. Our duties towards other creatures rise commensurate with the probability that they are sentient or conscious. Coeckelbergh and Gunkel (2014) similarly acknowledge the uncertainty intrinsic to efforts to determine the existence of internal states in other entities. Regardless of whether one is interested in assessing an animal's capacity for sentience (i.e., Singer) or being a subject-of-a-life (i.e., Regan), the "problem of other minds" (Dennett, 1981, p. 173) looms large. Therefore, because at present humans do not have the technological capability to directly observe what goes on in the minds of other beings, we can only render judgments about moral standing based on our perceptions of their externally available behaviors. Finally, Thompson (2019) concludes that "[t]here seems to be no empirical way to determine when different sentient beings are sufficiently alike to justify basic equality" (p. 23). As such, it is up to scientists to more rigorously define and identify concepts like autonomy. Research should equip judges with the kind of scientific clarity that can assist them in making decisions about the legal status of nonhumans.

Direct/indirect approaches offer an alternative to properties-based approaches to animal rights. The logic of these approaches reflects the Kantian view that the ethical treatment of animals is predicated not on a property they possess, but on the concern humans have for them or the apprehension that harming animals might lead to harm against humans (Zamir, 2007, p. 25). A few authors have pursued this line of argumentation, which balks at assigning independent moral status to animals. Two indicative examples are discussed below.

Writing at the turn of the 20th century, Salt (1900) urges that humans and animals are equals in kind, separated only by degrees of difference in terms of intelligence and sensibility. If we excuse the inhumane treatment of animals on the basis that they have lower quantities of these traits than normal humans do, criminals and savages, too, should suffer the same fate as that which befalls their animal kin.[10] While animals do not lay claim to full legal rights, we have duties towards them dictated by our senses of justice and reciprocity (i.e., the Golden Rule). The basis of ethics is this kinship of life, which does not depend upon duties being reciprocated among living beings.[11] In closing, Salt quotes Frederic Harrison (1904), who delivered the following remarks as part of a speech before the Humanitarian League, "[o]ur relation to the animals ... is part and parcel of our human morality ... Our duties towards our lower helpmates form part of our duties towards our fellow-beings" (p. 6). To summarize, our obligations towards animals are a function of our duties towards each other, not the attributes we share in common with nonhuman creatures.

Nearly a century later, Kelch (1999) pushes back against the idea that there is a single determinant of legal rights, arguing that human compassion should count among the reasons that such rights are extended to animals.[12] *Contra* both

Singer and Regan, Kelch seeks to find a place for emotions in moral theory. Far from functioning separately from reason, emotions reveal our values, focus our interests, and spur us to act, especially on behalf of others. Compassion in particular serves as the wellspring of our moral sensibilities. Our sense of compassion allows us to interpret the interests of animals, which are analogous to our own. While the manner in which compassion manifests itself in our relations with animals varies (i.e., attachment, kinship, or awe), the point remains that our emotions should be incorporated into the calculus we use to assess the extension of rights. This might be achieved by actually experiencing the kinds of conditions animals face in the laboratory, on the farm, or in the slaughterhouse. It is through the fusion of emotion and reason that humans can establish a moral basis for granting legal rights to animals.

These direct/indirect approaches are unabashedly anthropocentric—the treatment of animals depends on human duties towards others or human emotions. Duties emerge from our sense of justice, which Salt believes is universal. However, the myriad forms of animal ethics found in major world religions suggest otherwise. The notion of the kinship of life, on the other hand, provides a glimmer of daylight for disrupting the centrality of humans in an ethical system, hinting at the potential for radical equality among all living beings. Yet, the idea that duties towards animals are epiphenomenal to the duties we have towards other people unquestionably places the status of nonhuman creatures below that of humans. While including emotions in moral theory seems intuitively reasonable, this approach, too, assumes a kind of universalism belied by experience. Quite bluntly, not everyone feels compassionately towards animals. Some people may have limited interactions with animals and thus feel less attached to them (Hawkins et al., 2017). Others might experience only low levels of compassion, making them less inclined to act kindly towards humans, animals, or both. Still others may not consider themselves to be lacking compassion while using animals for pleasure, leisure, or sport in ways that cannot be seen as advancing the interests of nonhuman creatures (Tymowski, 2013). Therefore, despite evidence of its prevalence across cultures (Goetz et al., 2010), "advocating for compassion as a form of trans-species affinity is fraught with complication" (Chiew, 2014, p. 66).

As stated earlier, both the duties and emotions arguments are essentially anthropocentric. This is problematic because Western anthropocentrism is an "obsession" (Wu, 2014, p. 416) that in practice has elevated certain humans above others (Grear, 2013) and represents "the philosophical driving force behind ecological crises" (Hajjar Leib, 2011, p. 27). In the context of animals, anthropocentrism inevitably leads to prioritizing human interests over those of other creatures when concerns of the former run into conflict with concerns of the latter. As such, accepting that humans occupy a unique and superior position among living beings offers ethical cover for engaging in actions that may pose harm to other species.

A potential antidote to the conceptual, ontological, empirical, and moral issues associated with properties-based and direct/indirect approaches resides in relational approaches to animal rights. Analysts in this latter camp reject assigning moral status on the basis of human-like attributes, human interests, or human emotions. Instead, they promote a phenomenological perspective that draws moral conclusions based on how humans and other entities interact with animals. After all, animals are "relational beings who interact in diverse ways with the diverse other beings who share their home places" (Peterson, 2013, p. 12). Relying substantially on the work of Emmanuel Levinas (1906–1995), one pair of relational scholars argues that, for advocates of the approaches described above, moral status is determined prior to encounters with other beings. Therefore, the moral act occurs before interactions between entities ever take place. But Levinas posits that "the sequence is exactly the other way around: the ethical relationship, the exposure to the other, precedes the usual ontological decisions" (Coeckelbergh & Gunkel, 2014, p. 722). Reversing the Humean thesis, how we *ought* to treat an entity informs what it *is*.

Although Levinas did not write much about animals, he describes how the act of facing another entity fosters a "demand for recognition," a "moment that one becomes primordially aware of the ethical responsibility one bears towards the Other" (Crowe, 2008, p. 316). But *facing* is meant to be interpreted figuratively, not literally. There are many ways in which one can "face" an Other. Naming is one such mechanism. Absent being given a name, an Other "can be objectified, used, and even slaughtered since it is withdrawn from the sphere of moral considerability" (Coeckelbergh & Gunkel, 2014, p. 726). A pig is just potential pork until it is "Wilbur." Naming animals is an ancient practice permeated by power relations (Borkfelt, 2011). By giving names to animals, we distinguish them from others of their kind and signify their mortality (Derrida, 2002, pp. 378–379). Another mechanism by which animals face us is legal recognition. For instance, in 2019, the Punjab and Haryana High Court of India rendered a judgment finding that all animals are legal persons possessing the same rights and duties as humans (Malik, 2019).[13] Upon determining that animals share many qualities in common with humans, the High Court brought nonhuman creatures into the moral orbit through a legal declaration. In short, facing constitutes a range of practices through which responsibilities towards and rights of animals emerge.

Of course, one might criticize relational approaches on the grounds that they neither disentangle properties from animals involved in the act of facing nor extricate human interests from the moral calculus. Naming is something *humans* do to interpret the world around them and the entities within it. Names are thrust upon animals by *humans*. Legal recognition occurs because of *human* conflicts generated by *human* activities that are resolved through *human* institutions using *human* qualities as benchmarks. Therefore, it seems as though anthropocentrism inevitably invades virtually any effort to extend rights to nonhuman entities.

As demonstrated above, the literature on animal ethics has yet to reach a consensus regarding the optimal approach to deriving animal rights. However, this intellectual stalemate has resulted in a number of creative legal arguments conceived with the same intended objective in mind—protecting animals through existing systems. Several innovative legal pathways are reviewed below.

While agreeing with Bentham and Singer that the capacity to suffer is sufficient grounds for granting legal rights to animals, Sunstein (2000) focuses on the pragmatic issue concerning who has standing to represent the interests of animals in court. He offers that in the absence of legislative action that would expand the scope of persons recognized before the law, humans can file suit to protect animals under the guise of seeking redress for injuries to human interests. Using a quasi-Hohfeldian concept of rights, Sunstein also argues that laws against animal cruelty are functionally equivalent to legal rights because they provide animals with an immunity or guarantee against ill-treatment.

In a subsequent article, Leslie and Sunstein (2007) expand upon this latter point. The authors recognize that animal cruelty laws feature certain limitations that inhibit their ability to fully secure rights-like protection for animals. First, the only route to enforcing animal-welfare statutes is through public prosecution of offenders, and there are practical reasons why such violations are not likely to result in legal action. Second, animal cruelty laws are severely under-inclusive. That is, they do not pertain to the treatment of animals used in laboratory testing or food production. Realizing these weaknesses, the authors propose an alternative means of safeguarding animals—disclosure requirements (i.e., food labeling). Food labels constitute a hallmark of the "informational turn in food politics" in which responsibility associated with consumption is shifted from public to private institutions (Frohlich, 2016, p. 162). Labeling food products could signal compliance with animal welfare standards, providing consumers with information about a producer's treatment of animals. This in turn would generate market pressure that might inspire a race to the top among producers. As a concrete example, "a label might disclose the frequency (or absence) of bruises, broken wings, and birds that are dead on arrival at the processing plant, all of which can result from rough handling" (Leslie & Sunstein, 2007, p. 134). To recap, the authors suggest that laws against animal cruelty serve the same function as legal rights, although because of issues with their enforcement, market-based mechanisms such as labeling provide an alternative route to achieving the same objective.

While this kind of innovative thinking is a welcome addition to the animal rights literature, the emphasis on animal welfare, as opposed to animal rights, proves problematic for several reasons. First, the pervasiveness and strength of animal welfare laws vary widely across jurisdictions, meaning that animals will enjoy more or less robust protection depending on where they live (Park & Valentino, 2019). For instance, according to the Animal Legal Defense Fund (2020), Illinois (rated the best overall state) affords animals far greater legal protection than does Mississippi (rated the worst overall state). Clearly, animal welfare is distributed on a highly unequal basis because it means different things in different places. Second, rights provide animals with "a stronger and more sustainable protection" than that which is available through anticruelty laws (Peters, 2016, p. 49). Such laws place animal welfare in direct competition against other interests, with those held by humans typically being prioritized at the expense of animals, even when they are trivial by comparison (Peters, 2018, p. 356). Third, as Leslie and Sunstein note, anticruelty statutes don't protect the majority of animals, so perhaps a more effective means of curtailing harm to animals would be to shift the focus to limiting human entitlements by "sacrificing" the human right to kill animals (Bryant,

2008, p. 256). Fourth, animal cruelty laws are essentially anthropocentric, as they were conceived "to protect human sensibilities," not respect the interests of animals themselves (Cullinan, 2003, p. 77). Given the ubiquity of animal welfare laws, one way of building the case for their insufficiency is by engaging in a "litigating to lose" strategy whereby suits are filed under the assumption that they will be unsuccessful but generate significant media attention, highlighting the need for greater protections (Vayr, 2017, p. 873). Despite the potential upside of costly legal efforts predestined to fail, as long as animals are considered human property and not independent beings with interests of their own, they will be denied entry into the moral circle and unable to access its attendant benefits.

Two recent contributions to legal scholarship on animal rights present strategies designed to overcome the persistence of the personal property paradigm, albeit from different angles. Abate and Crowe (2017), like Wise before them, proceed inductively, looking to the decisions of jurists for inspiration. Recognizing that while owned things are considered property and thus legal objects (Cullinan, 2003, p. 77), there already exists legal precedent for treating a range of nonhuman entities as legal subjects entitled to legal personhood. The central inquiry therefore revolves around whether or not the same logic used to determine that nonhuman entities can be considered legal subjects applies to animals as well. To begin, the authors describe animals as members of "the community of the voiceless," a group of legal subjects that "cannot assert and vindicate their interests without legal personhood recognition and 'guardians' to act on their behalf" (Abate & Crowe, 2017, p. 71, n114). Among the constituents of this community, which includes artificial intelligence (AI) and future generations, natural resources such as ecosystems, mountains, and rivers serve as the closest analog to animals. Natural resources have been afforded legal personhood in countries around the world on the basis of their inherent value. Both natural resources and animals possess inherent value "as critical components of our ecosystem and as entities that hold deep cultural, spiritual, and emotional value in our lives" (Abate & Crowe, 2017, p. 77). In addition, humans inarguably enjoy apex legal status, and animals are more similar to humans than are elements of nature because they possess sentience and the capacity to suffer. Therefore, there is no reason why animals should not also qualify for legal personhood and its accompanying rights.[14]

Bradshaw (2018) seeks to shift the focus of the animal rights debate away from those beings humans encounter in farms, homes, and zoos, and towards those animals affected by unsustainable land practices—wildlife on the land and in the sea. The problem, as she sees it, is not that such animals are considered human property. After all, wildlife creatures are not owned by anyone. Rather, the issue is that these animals are excluded from the property rights regime; that is, they are deprived of legal authority over the lands and seas on/in which they reside. Therefore, Bradshaw proposes a novel solution—granting property rights to animals. While wildlife would possess these rights, these legal protections would be administered by "human representatives vested with a fiduciary duty to oversee the intergenerational wellbeing of all creatures within an animal-owned ecosystem" (Bradshaw, 2018, p. 813). This trusteeship model comports with existing legal institutions and endeavors to protect the interests of animals often neglected by the animal rights movement, whose homes are constantly threatened by human activity.

While both of the aforementioned legal strategies—comparing animals to natural resources and granting property rights to wildlife—offer innovative, practical solutions to the property paradigm problem, they are not without drawbacks. First, Abate and Crowe's natural-resources analogy leans on the notion of inherent value drawn from environmental ethics. While they do not elaborate on what they mean by inherent value, the phrase speaks to an ethical orientation that is at least "weakly anthropocentric" (Norton, 1984, p. 134). Briefly, when humans ascribe value to nature, it becomes an object worthy of protection. This view has been critiqued extensively in the literature (i.e., Weston, 1985; Morito, 2003), with one scholar going so far as to assert that "there is no basis for this claim that is not rooted in either theism or some other form of arbitrary value judgment that, in a materialistic worldview, has no basis outside of the believer's mind" (Snyder, 2017, p. 59). Ironically, Callicot (1995) compares intrinsic value[15] to consciousness, arguing that we know that both exist because we experience them internally. However, as detailed in Chapter Two, consciousness is an empirically elusive quality whose presence in other entities remains inaccessible. Therefore, while animals might be akin to natural resources, the analogy suffers from a lack of clarity regarding what inherent value means and how it applies in both cases. The matter is complicated further when Abate and Crowe draw a parallel between humans and animals given their common possession of sentience. As discussed in Chapter Two, sentience is a property that is alleged to qualify an entity for psychological personhood. Thus, the authors' reliance on sentience makes their legal argument philosophically committed to the flawed properties-based approach advanced by Singer. Second, while Bradshaw seeks to offer legal shelter to wildlife in a way that does not require demonstrating how animals are similar to humans, her approach restricts the kinds of animals eligible for protection. As the author herself admits, a revised property rights regime would not address the plight of animals found in farms, laboratories, or zoos. To Bradshaw's credit, she maintains that granting property rights to wildlife should be viewed as a strategy complementary to, rather than replacing, other efforts to expand rights to animals. As such, animal property rights represent both a radical departure from current legalized forms of animal protection and an incremental step towards the ultimate goal of animal liberation.

Philosophers writing on the moral status of intelligent machines have paid great attention to the literature on animal ethics. In particular, they have assessed the extent to which theories of animal rights might apply to technological beings. Identifying key similarities and differences between these types of entities might help inform responses to the machine question, refining the boundaries of the moral circle in the process. As Gunkel (2012) points out, "the extension of moral rights to animals would, in order to be both logically and morally consistent, need to take seriously the machine as a similar kind of moral patient" (p. 81). In the space here, I summarize some of the main contributions in this area according to two related lines of inquiry.

First, are animals and machines sufficiently similar such that useful comparisons can be drawn? While Descartes found animals to be much like machines (P. Harrison, 1992), recent scholarship suggests an answer in the negative. For Calverley (2006), the fact that technological forms are created directly and deliberately by humans, and considering that animals possess autonomy and consciousness, an analogy between animals and machines might be inapt. Gunkel (2012), too, casts doubt on the utility of attempting to establish an animal–machine parallel. Citing philosophical work on animal rights, he observes that the presence or absence of sentience appears to divide entities worthy of moral consideration from those that are not. Referring back to the problem of other minds, he notes that "[w] e are ultimately unable to decide whether a thing—anything animate, inanimate, or otherwise—that appears to feel pain or exhibits some other kind of inner state has or does not have such an experience in itself" (Gunkel, 2012, p. 100). Thus far, the differences between animals and machines seem to make a helpful analogy unlikely.

Second, despite the dissimilarities between the two types of entities, could theory on animal rights provide a basis for robot rights? On this question, responses vary. Some squarely reject the animal rights model. For McGrath (2011), the lack of consensus among animal rights theorists suggests it is "perhaps best to not approach the muddy waters of robotic rights by way of the almost equally cloudy waters of animal rights" (p. 135). Torrance (2013) explains that animal rights extend from a biocentric environmental ethic that emphasizes an entity's capacity for moral experience and sentience. Artificial agents that do not possess these qualities would not have a serious claim to moral interests, and thus not be entitled to moral rights. Hogan (2017) observes that intelligent machines serve as a "limiting case, whose very exclusion from the ethical realm is necessary for animal rights discourse" (p. 32). In other words, robots function as a device that strengthens the argument in favor of animal rights when conducting an inventory of morally significant qualities identified among various entities. Therefore, robots clarify the case for animal rights. Michalczak (2017) argues nearly the opposite. Software agents, he maintains, are more likely than animals to be recognized as legal subjects because their ethical status rests on pragmatic, anthropocentric considerations amenable to existing human institutions. Whereas the legal subjectivity of animals often depends on the demonstration of certain properties and can pit human interests against those of nonhuman creatures, the legal subjectivity of intelligent machines advances human interests similar to the way in which the personhood of corporations and ships was manufactured to resolve human conflicts.[16] More concretely, legal subjectivity could be a tool useful in the adjudication of disputes involving autonomous weapon systems and algorithmic trading programs, especially in cases where establishing the party responsible for causing an alleged harm proves troublesome. Thus, an ethical framework that leans into human interests suggests that robot rights are unlikely to follow the animal rights model, and intelligent machines might therefore be more likely than animals to obtain legal status and legal rights.

Others adopt a more conditional stance. Calverley (2006) holds that if an intelligent machine demonstrates possession of attributes found to warrant extending moral consideration to animals, humans would have to treat such an entity as a being with value in order to avoid charges of speciesism. Interestingly, Singer has offered commentary on the applicability of animal rights theory to the realm of technology. Writing with Agata Sagan, Singer (2009) posits that if robots were programmed in such a way that they developed consciousness spontaneously, as opposed to being deliberately designed to mimic consciousness, we would be compelled to seriously entertain their case for rights. Finally, Marx and Tiefensee (2015) draw a distinction between (domesticated) animals and robots, concluding that while the former possess rights but not moral agency, the latter could in theory possess both.[17] In order to qualify for rights, an entity needs to be sentient and have interests worth protecting. If there is no "moral difference whether frustration or pain is felt the human or the robot way," intelligent machines might be said to have interests and thus rights (Marx & Tiefensee, 2015, p. 86). Advocates of strong AI contend that sentience and interests are theoretically possible, and some robots already behave in ways that suggest a functional similarity to human actions.

In brief, philosophers have looked to animals and theories of animal rights for guidance on the machine question. But they have found animals to be vastly different from robots in morally meaningful ways and, perhaps unsurprisingly, determined that animal rights offer robot rights insights that apply mostly under speculative conditions. Therefore, the prospects of arriving at a theory of robot rights through literature on animal ethics seems quite remote, at least for now.

The limited degree to which animal rights theories can be usefully translated to a technological context stems from problems endemic to extant animal ethics scholarship. A few main weaknesses are discussed here. First, setting aside for a moment the obvious differences between philosophical and legal arguments, much of the writing on animal ethics concentrates intensively on how requisite properties determine rights while almost entirely neglecting the conceptual importance of personhood (Aaltola, 2008).[18] This is surprising not only because many scholars have painstakingly sketched out paths leading from certain properties to various types of personhood, but also because "[i]n law, personhood is a precondition for holding rights" (Peters, 2018, p. 357). To wit, neither Singer's *Animal Liberation* nor Regan's *The Case for Animal Rights* mention the word "personhood" even *once*.[19] Second, the explicitly non-anthropocentric, individualistic orientation of major animal rights theories implicitly accepts anthropocentric, dualistic thinking while "ignor[ing] the importance of the web of life within which all species are situated" (Bryant, 2008, p. 266). This Cartesian mindset tends to produce animal rights theories that are under-inclusive (i.e., pertaining only to animals in captivity, domestic pets, or wildlife) and thus of limited utility as ethical frameworks. A more robust ethic would "[account] for and [do] justice to all of nature, including individual nonhuman subjects and entire ecosystems, as well as the many natural places and entities in between" (Peterson, 2013, p. 141). Third, and most importantly, animal rights theories rely on the demonstration of one or more ontological properties that serve to validate the presence of interests held independently by animals. The evidence used to suggest possession of these properties ranges from the empirical to the intuitive to the emotional to the jurisprudential. However, as of yet, no one has been able to identify a single attribute or hierarchy of attributes capable of fostering consensus among animal rights theorists. The indeterminacy of the argument from marginal cases illustrates this in spades. These critiques suggest that theories of animal rights might prove unhelpful to the task of assessing the extent to which other kinds of nonhumans, be they organic or technological, might be eligible for moral or legal rights.

In order to clarify the differences between the four approaches to animal rights described above, I briefly remark on how each of them respond to several important questions. First, *what is the basis for moral or legal consideration?* Properties-based approaches put forth one or more attributes (i.e., consciousness, intentionality, sentience, etc.) deemed necessary or sufficient for the extension of moral consideration and moral rights. Direct/indirect approaches translate human interests or human emotions into moral concern for animals. Relational approaches depend on the nature of interactions between humans and animals, not preconceived notions about the kinds of properties that qualify other entities for moral consideration. Legal approaches offer a mix of justifications for legal consideration, including animal suffering (Sunstein, 2000; Leslie & Sunstein, 2007); inherent value, the capacity to suffer, and sentience (Abate & Crowe, 2017); and property interests (Bradshaw, 2018). Second, *what is the ontological orientation?* All of the approaches are individualist with the potential exception of relational approaches, whose orientation could be individualist or holist depending on the ontological scope of the encounter (i.e., human-and-animal, human-in-ecosystem, etc.).

Third, *what is the extent of human centrality?* Properties-based approaches tend to be non-anthropocentric in the sense that they afford moral rights to entities other than humans, but the properties themselves privilege human characteristics over those of animals. Direct/indirect approaches are also conceptually non-anthropocentric because they accept the possibility of animal rights, but the only way to arrive at that destination is through human interests or feelings. Relational approaches are more non-anthropocentric than the other two types given that they remain open to rights for any entities humans may encounter and they do not prescribe ethical guidelines in advance of interactions with nonhumans, but the mechanisms by which animals are recognized as moral patients require human intervention. Legal approaches entail varying degrees of human emphasis. While animal welfare is fairly anthropocentric due to its lack of commitment to animal rights and reliance on human sentimentality, a belief in the inherent value of non-human creatures that results in their obtaining legal personhood and legal rights is decidedly non-anthropocentric, despite the fact that their case is bolstered by the similarities they share with humans.

Weaving the threads of philosophy and law together, how might animal ethics serve as a foundation for extending rights to nonhuman inorganic entities such as intelligent machines? First, assuming that philosophers can eventually agree on the single property or set of properties that necessitate(s) moral consideration, we would need to empirically demonstrate the existence of said property/properties in nonhuman inorganic entities in order to validate their eligibility for moral rights. With respect to the properties-based approach, autonomy, consciousness, or sentience would be the most likely candidates. At present we are far off from being able to recreate these properties technologically or determine their presence scientifically. However, there do exist potential work-arounds available in the interim. Following Chan (2011), we could utilize a probabilistic model to determine the level of uncertainty we have regarding an entity's possession of a given trait and calibrate the strength of our moral consideration accordingly. Moral rights could be extended where the degree of uncertainty falls below a certain threshold. Wise (2013) applies a similar logic with his continuous measurement of practical autonomy that results in four categories of ascending legal consideration. However, for both of these techniques, we would still need to know which property or properties is/are the most philosophically defensible for the purposes of analyzing the level of uncertainty or determining a numerical value. The level of uncertainty or number ranging from 0.0 to 1.0 obtained would dictate first whether or not an entity qualifies for moral/psychological personhood or legal personhood; then, whether it is a moral agent/patient or legal object/subject; and finally, whether it is entitled to moral or legal rights.

Second, in line with Kelch (1999), nonhuman inorganic entities could be afforded moral consideration based on the level of emotional attachment humans have towards them. The stronger the emotional attachment, the greater the case for moral concern. Practically speaking, this strategy might be difficult to implement because each relationship would need to be evaluated independently, and the depth of compassion that humans feel towards such entities will vary widely.[20] But that isn't the only area where consistency might present a problem. For example, if someone were to love their Amazon Alexa but hate all dogs, would violence against the latter, but not the former, be permissible? Therefore, emotion might be a necessary, but not sufficient, condition for a logically coherent approach to determining moral consideration.

Third, observing that intelligent machines constitute voiceless legal subjects (Abate & Crowe, 2017), one way to extend protection to members of this misfit group involves recognizing their value to ecosystems and humans. As natural resources have already been determined to possess legal personhood on this basis, in principle there is no reason why robots could not also be afforded the same distinction. Although this argument will be developed more in Chapter Five, for the moment it suffices to say that under certain conditions robots can also hold value due to the roles they play in social-ecological systems, which qualify them for legal personhood and thus legal rights. Of course, this maneuver requires adoption of a holistic conception of the environment that collapses the Cartesian nature/culture divide.

Fourth, facing artefactual entities through the mechanisms of naming and legal recognition presents another means of rendering others with whom humans interact worthy of moral consideration. Indeed, humans already have a legacy of engaging with technology in these relational ways. The human tendency to anthropomorphize non-living beings is well documented across cultures (Boyer,

1996),[21] and governments have begun discussing the legal status of technological agents (European Parliament, 2017). It seems like only a matter of time before Michalczak's (2017) intuition about the legal subjectivity of intelligent machines comes to fruition, as the practical benefits of bringing technological entities into the legal domain will come to outweigh the costs to human exceptionalism. The idea that moral status might be determined by legal recognition seems confusing at first unless we refer back to the conceptual mapping exercise from the previous chapter, in which facing is considered a mechanism by which relational person-hood and thus both legal and moral personhoods are demonstrated.

To conclude, each group of approaches presents a different benchmark for moral or legal consideration—possession of a property or combination of proper-ties, the activation of human interests or emotions, the extent of human encoun-ters with and recognition of other entities, or a medley of legal strategies ranging from the abstract (i.e., inherent value) to the concrete (i.e., food labeling). Each contain their own strengths and weaknesses. Properties-based approaches offer conceptual parsimony, but suffer from empirical and philosophical indetermi-nacy. Direct/indirect approaches rely on more readily verifiable human condi-tions, but overlook the moral or legal subject itself and obscure the variability among humans that could lead to incoherent standards. Relational approaches privilege real experiences with nonhuman entities, but the tenor of those encoun-ters is shaped by some of the very characteristics that these approaches seek to avoid. Legal approaches provide a range of solutions steeped in pragmatism, but their diversity often leads to under-inclusiveness.

Perhaps not surprisingly, none of the approaches described above put forth the argument that animals should have rights according to will theory. Instead, it is the interests of nonhuman creatures or their human guardians that lead to animal rights.[22] As mentioned previously, it is a curious feature of animal ethics literature that personhoods and statuses are virtually absent from the debate. The previous chapter showed how forms of personhood are integral to the determination of moral or legal status and then moral or legal rights. Finally, while some doubt the applicability of animal rights models to technological entities, a few of the arguments reviewed here suggest potential pathways worthy of further explora-tion. The next section evaluates how animal rights have fared in courts around the world, with a view towards understanding which approaches appear to have trac-tion and how the successes and failures of animal rights advocacy might elucidate the potential for robot rights.

Animal rights in court

Although modern philosophy on animal rights emerged in the 1970s and 1980s (Cupp, 2017, p. 3), animals have stood trial for alleged offenses since the medi-eval period (Girgen, 2003). In these early cases, animals were tried as legal per-sons in criminal proceedings. Over the past decade, however, there has been a dramatic rise in cases seeking to advance the rights of animals themselves. In these instances, animal rights theories have had their practical utility assessed by jurists across the globe. This section of the chapter reviews several recent cases on animal rights in the United States, India, Argentina, and Colombia. The purpose is to evaluate the extent to which the approaches to animal rights detailed above have proven persuasive in courts of law. While this comparative exercise is by no means exhaustive, the cases selected for analysis represent major developments in the field of animal rights. In the course of this analysis I pay special attention to the use of concepts such as personhood and status in an effort to derive insights that might be applicable to the question of rights for nonhuman inorganic entities.

In *People ex rel. Nonhuman Rights Project, Inc. v. Lavery* (Lavery I),[23] the NhRP sought to appeal a New York State Supreme Court judgment denying the group's application to initiate a *habeas corpus* proceeding on behalf of Tommy, a chimpanzee, whom they argued was being unlawfully detained. According to NhRP, Tommy was "living along in a cage in a shed on a used trailer lot along Route 30 in Gloversville, New York" (Nonhuman Rights Project, n.d.). The case revolved around the question of whether or not Tommy qualified as a "person" capable of possessing an "interest in personal autonomy and freedom from unlawful detention."[24] The NhRP argued that chimpanzees exhibit cognitive abilities similar to those found in humans, including autonomy and self-awareness. Therefore, they should be considered persons eligible for the protections available under the writ of *habeas corpus*. The court declined to expand the definition of a legal person to include animals, finding that "animals have never been considered persons for the purposes of habeas corpus relief."[25]

In rendering its decision, the court examined the concepts of person and legal personhood, and evaluated the extent to which they might apply to nonhumans. To clarify these terms, the judges looked to Black's Law Dictionary, which defined "person" as both "[a] human being" and "[a]n entity (such as a corporation) that is recognized by law as having the rights and duties [of] a human being" (emphasis omitted).[26] Finding the possession of rights and duties central to the description of a person and thus required for the demonstration of legal personhood, the court then cited domestic case law in the U.S. supporting this connection between incidents and personhood. The specific aspect in which chimpanzees fell short and corporations proved defensible was legal accountability. For the judges, the "incapability to bear any legal responsibilities and societal duties" made the extension of legal personhood and thus legal rights to chimps "inappropriate."[27] Further, the court held that New York's animal cruelty laws offered sufficient safeguards for animals, and encouraged the petitioners to press the legislature to enhance the legal protections available to chimpanzees. Following several unsuccessful attempts to appeal the court's decision and the filing of a second *habeas corpus* petition,[28] which was also denied, appealed, and denied again, the case came to an end in 2018, when the New York Court of Appeals denied a motion to allow NhRP to appeal.

Aside from the facts that the court appears to have conflated natural persons with artificial persons, referring to both simply as kinds of "persons,"[29] and that "persons" and "legal personhood" are used interchangeably,[30] a small but meaningful definitional issue led to a potentially erroneous conclusion about Tommy's legal status. As attorney for the petitioner Elizabeth Stein indicated in her letter to the Clerk of the Court in the First Department's Appellate Division, the definition of "person" featured in Black's Law Dictionary (7th edition) misquoted two of its supporting sources, which actually promoted the idea that a "person" entails a being who has rights *or* duties (Stein, 2017). Although this was an important oversight, the case failed to muster the judicial support necessary to proceed.

The tortuous path that the *Lavery I* case traveled along the way to validating the status quo illuminates the gap between legal arguments advanced by animal rights advocates and the jurists charged with deciding their merits. The NhRP utilized a properties-based approach to extending rights to animals that would have such nonhuman entities qualify for moral and/or psychological personhoods, which serve as a prerequisite for legal personhood. But the New York State Supreme Court rebuffed this argument, observing a dearth of precedent finding that animals are persons, relying on a definition of legal personhood that requires entities to be capable of exercising both rights and duties, and finding that chimpanzees do not possess either of these capabilities.

Two key insights from this judgment foretell the likelihood that courts might extend rights to robots. First, at least in the U.S. context, the properties-based approach to animal rights is insufficiently persuasive for obtaining judgments in favor of protecting nonhuman entities on the basis of their individual rights. Short of demonstrating that intelligent machines can effectively execute duties and be held legally accountable for their actions, they are unlikely to be deemed persons under the law. Pursuit of this objective is frustrated by the fact that legal scholars and technologists have yet to come to an agreement regarding how to best address the myriad legal accountability issues intrinsic to AI (i.e., Lehmann et al., 2003; Hallevy, 2010; Doshi-Velez & Kortz, 2017). As the court indicated, corporations, unlike chimpanzees, are associations of humans that bear legal duties. Only those entities in possession of moral agency and the capacity for societal responsibility can exercise duties and thus be considered legal persons entitled to rights.

Second, if the category of legal persons comprises only those entities capable of duties, responsibility, and accountability, then, absent scientific evidence suggesting technological beings are capable of these things, the only way they can obtain legal rights would be through one of the theories extending legal personhood to corporations. For instance, a robot rights group could form an association on the grounds that they possess a common interest in the welfare of AI, with all of the attendant privileges, responsibilities, and liabilities contained therein. Then the robot could be considered a legal person whose accountability reaches back to the humans who incorporated on its behalf. However, this strategy would likely fail to protect all intelligent machines, and it would depend on a property rights–like model that might invite controversy among animal liberationists. An alternative would be to classify robots as legal minors and designate humans as persons in *loco parentis* who serve as their guardians. Such humans could then be considered responsible for the actions of technological entities. This approach has already been deployed in an Indian case involving the personhood of rivers (O'Donnell, 2018, pp. 142–143).[31]

In *Karnail Singh and others v. State of Haryana* (Karnail Singh),[32] several truck drivers and conductors had previously been found guilty of violating the Punjab Prohibition of Cow Slaughter Act for exporting a total of 29 cows from the State of Haryana to the State of Uttar Pradesh. The petitioners appealed the conviction, but they received only a reduced sentence and the ruling was not overturned. Thereafter, the Punjab and Haryana High Court agreed to review the revised petition against the lower court's verdict. After quickly dispensing with the judgment that upheld the conviction and sentenced the petitioners to time served, the court turned to the issue of animal cruelty.

The sprawling, 104-page decision steadily builds towards an outcome in favor of extending legal personhood and legal rights to animals by drawing upon Indian regulations and case law, American jurisprudence, legal scholarship, non-Western religious doctrine, quotes from renowned Eastern leaders, and the Indian Constitution. The evidence might be summarized as advancing the following arguments: (1) some nonhuman entities have already been classified as juristic or legal persons under Indian or American law; (2) scientific evidence shows that animals possess certain human-like traits; (3) Eastern religions support the personhood of animals and the need to be compassionate towards them; (4) the Indian Supreme Court has interpreted the fundamental right to life as applying to all species, including animals;[33] (5) the law must respond to ecological destruction by creating new instruments designed to protect the environment; and (6) the Indian Constitution articulates a duty to have compassion for all living creatures,[34] and mandates that the state organize agricultural practices[35] and safeguard the environment, which includes animals.[36] The ruling closes with a sweeping declaration:

> The entire animal kingdom including avian and aquatic are declared as legal entities having a distinct persona with corresponding rights, duties and liabilities of a living person. All the citizens throughout the State of Haryana are hereby declared persons in loco parentis as the human face for the welfare/ protection of animals. "Live and let live."[37]

The *Karnail Singh* decision relies on several of the approaches to animal rights described in the previous section. First, following an analysis of legal personhood in the American and British contexts, the judges give credence to the properties-based approach by citing David Boyd's *The Rights of Nature* (2017), which offers evidence that animals possess emotions, intelligence, self-awareness, and altruism. Second, the Court advances a direct/indirect approach through its attention to the principle of *ahiṃsā* and the way in which practicing compassion towards others reflects compassion towards oneself. Another direct/indirect approach speaks more broadly to how the concept of juristic persons is used to advance the needs of society. Finally, the ruling applies two legal approaches to the question of animal rights. The first involves animal welfare laws, which the Indian Supreme Court has interpreted as guaranteeing animals five freedoms: "(i) freedom from hunger, thirst and malnutrition; (ii) freedom from fear and distress; (iii) freedom from physical and thermal discomfort; (iv) freedom from pain, injury and disease;

and (v) freedom to express normal patterns of behavior."[38] The second recognizes the inherent value or intrinsic worth of animals, which emerges from an ecocentric ethical orientation towards nature.[39] Recognizing that animals have a right to life, they cannot be considered mere property[40] and their existence is not predicated on the instrumental value humans assign them.

To be sure, the judgment rendered in *Karnail Singh* presents a multi-pronged justification for animal rights that divines inspiration from different jurisdictions. But how might this landmark decision inform the debate over rights for AI? The answer is that the extension of rights depends on the sophistication of the technology, the manner in which humans interact with other entities, the potential benefits obtained by treating nonhuman inorganic beings as legal persons, and the ontological scope of the moral circle. It remains unclear whether the factors enumerated above serve as conditions sufficient or necessary for the enjoyment of legal rights. The Indian approach could be a multi-criterial anomaly with little prospect of translating intelligibly to other legal systems. However, the court's reasoning applied in *Karnail Singh* does suggest that responses to the question of legal rights for robots will probably require an examination of several factors, including characteristics of the legal subject; domestic and foreign jurisprudence; local, national, and international cultural contexts; the current capacity of law to address existential threats; and the ontological boundaries of the environment.

In *Orangutana Sandra s/ Recurso de Casación s/Habeas Corpus* (Sandra I),[41] the Association of Professional Lawyers for Animal Rights (AFADA) sought relief in the form of a writ of *habeas corpus* for Sandra, an orangutan born in Germany and living in the Buenos Aires Zoo in Argentina. AFADA claimed that Sandra was being deprived of her freedom and subject to conditions imperiling her health, and called for her immediate transfer to a primate sanctuary in Brazil (de Baggis, 2015, p. 2). After the argument was rejected by two inferior criminal courts and twice appealed, at the end of 2014 the Federal Chamber of Criminal Cassation ruled on the issue, remanding the case to the criminal court of Buenos Aires for jurisdictional reasons (Wise, 2015a). In its terse 2-1 decision, the Chamber issued the following provocative statement:

> That based on a dynamic rather than a static interpretation of the law, it is necessary to recognize the animal as a subject of rights, because non-human beings (animals) are entitled to rights, and therefore their protection is required by the corresponding jurisprudence (citations omitted).[42]

This part of the ruling gained international media attention, resulting in hyperbolic claims about the far-reaching implications of the decision.[43] However, as Wise (2015b) points out, "the court had neither issued a writ of habeas corpus, nor granted Sandra personhood for any purpose, nor ordered her moved to a sanctuary." Further, the statement alleging that animals have rights amounted to dicta, which refers to propositions used in the course of legal reasoning that are not actually decided, are not based upon the facts of the case, or do not lead to the ultimate judgment (Abramowicz & Stearns, 2005, p. 1065).

Despite these issues, it is worth examining the underlying rationale the judges applied to reach this controversial conclusion, which can be gleaned from the internal citations provided at the end of the statement. Both of the sources referenced were penned by Eugenio Raúl Zaffaroni, a former member of Argentina's Supreme Court of Justice and current judge on the Inter-American Court of Human Rights.[44] In the first work, Zaffaroni et al. (2002) argue that direct/indirect approaches to animal rights have proven to be flawed, and that nonhuman subjects (i.e., animals) provide socially valuable "legal assets" (*bienes jurídicos*) such as the preservation and conservation of different species (p. 493).[45] In the second work, Zaffaroni (2011) contends that the legal asset in animal cruelty cases is the "right of the animal itself not to be the object of human cruelty" (p. 54).[46] He also pushes back against the argument that animals do not have rights because they are incapable of demanding them (i.e., they lack legal capacity and are ineligible for rights under will theory), citing marginal cases of humans who are similarly ill-equipped to partake in legal matters. Therefore, despite their brevity, the judges in this case had in fact supported their dynamic assertion of animal rights using a legal approach supplied by one of Argentina's famed jurists.

Sandra's legal status left unresolved, AFADA pursued an *acción de amparo*, a legal mechanism conceived to protect basic rights by providing "a quick solution to urgent circumstances" (Thompson, 2019, p. 16).[47] On October 21, 2015, Judge Elena Amanda Liberatori issued a ruling on the case (Sandra II).[48] Agreeing with the precedent set by the previous court in 2014, Judge Liberatori found no reason why Sandra should not be considered a nonhuman person[49] and subject of rights. Her decision focused on the application of Argentine Civil Code, which she interpreted in light of 2015 reforms to the Civil Code in France. In January of that year, the French Parliament voted to change the legal status of domestic pets and wild animals held in captivity from "movable property" to "living beings gifted sentience" (Hervy, 2015).[50] Judge Liberatori argued that the French development should be used to inform the interpretation of Argentina's own Civil Code, thus designating Sandra a nonhuman person on the basis of her sentience, which was demonstrated by a panel of experts.

In support of applying new categorizations in the context of law, she cited Zaffaroni's discussion of the rights of nature, specifically those found in the Ecuadorian Constitution. Ecuador's governing charter, and to a lesser extent Bolivia's, invoke an "andean ecology" (Bandieri, 2015, p. 33) that recognizes Earth as a subject of rights. The larger point that Judge Liberatori sought to make about legal subjects reflects the jurisprudential outlook of legal positivism—categories are social constructions that are inherently dynamic, not natural and static. They change according to the evolving social context. Furthermore, she held that "it is necessary to denature and problematize the way in which one thinks on a daily basis, since this way of thinking has been socially and historically constructed for centuries and can enclose relations of domination and inequality."[51] Thus, interrogating how humans have classified reality allows us to reveal power relations and combat the subjugation of all living beings.

After Judge Liberatori's ruling, Sandra's situation remained mired in uncertainty. The Buenos Aires Zoo closed and turned into an eco-park, and the issue again came before the judge, who temporarily rejected AFADA's request to have Sandra transferred to a primate sanctuary (GAP Project, 2017). Due to the dearth of accredited sanctuaries in Argentina, Judge Liberatori requested that Sandra be relocated to the Center for Great Apes (CGA) in Wauchula, Florida (Elassar, 2019). After a month-long quarantine at the Sedgwick County Zoo in Kansas, Sandra was moved to CGA, where she has resided since 2019.

Both of the decisions regarding Sandra touch upon a variety of the approaches to animal rights discussed in this chapter. In the former case, direct/indirect approaches were implicitly rejected in favor of non-anthropocentric legal approaches viewed through the lens of legal positivism, with a nod towards properties-based approaches via the argument from marginal cases. In the latter case, a legal approach was informed by a properties-based approach, presented through an explicitly non-anthropocentric (though still largely individualist) application of Indigenous ideas, legal positivism, and critical theory. In summary, both cases weave several approaches together, proffering legal arguments for animal rights that reinterpret the concept of legal subject using combinations of pragmatism, dynamism, science, and external ideas.

The decisions rendered in both Sandra cases provide uneven support for extending rights to intelligent machines. In the first case, while a dynamic interpretation of the law might enable the possibility of rights for artefactual entities, the logic employed by the court seems to reverse the order of concepts described in Chapter Two; here, rights lead to the determination of legal status (i.e., object or subject). This process entails demonstrating that an entity possesses legal rights prior to assessing its legal status. Given that the court did not explain how animals were adjudged to possess rights in the first place, the only clues available come from the sources cited, which suggest that finding animals to be legal subjects serves primarily to benefit those creatures and secondarily to confer social value to humans. In the second case, a similar preference for disrupting the static nature of law shines through, leaving the door open to new ideas and potentially new legal subjects. Revisiting old ways of legal thinking is necessary in order to address power imbalances and inequality. Further, a willingness to look outside one's own jurisdiction for inspiration via "transjudicial communication" (Slaughter, 1994, p. 101) or "legal transplantation" (Schauer, 2000) could very well result in alterations to existing legal paradigms. Therefore, at least in theory, extant legal categories could be altered in the face of changing circumstances. However, the extent to which this dynamism might facilitate the rights of nonhuman non-animal entities remains unclear.

In a 2017 decision from Colombia's Supreme Court of Justice,[52] Judge Luis Armando Tolosa Villabona granted a writ of *habeas corpus* to a spectacled bear named Chucho, overturning a ruling by the Civil Family Tribunal of the Superior Court of the Judicial District of Manizales. Luis Gómez, the lawyer and petitioner acting on Chucho's behalf, argued that in transferring the bear from his previous home at the Río Blanco Nature Reserve to the Barranquilla Zoo, the

Corporación Autónoma Regional de Caldas "had deprived Chucho of his right to liberty, severely compromising his physical and emotional well-being" (Choplin, 2019). Further, Gómez maintained, the endangered bear was currently being held under conditions tantamount to "permanent imprisonment" (Stucki & Herrera, 2017). As such, he sought relief for Chucho in the form of permanent relocation to the La Planada Natural Reserve in the state of Nariño.

In the process of evaluating the petitioner's claim, Judge Tolosa expounded a critical, context-specific legal argument for animal rights steeped in environmental ethics, legal theory, domestic law, and international soft law. He began by laying the blame for ecological destruction at the feet of Enlightenment philosophers, whose rationalism, individualism, and capitalism enabled humanity's rapacious exploitation of the environment. Humans have become conquerors, "immoderate and irresponsible beings," who have "ignor[ed] that it is they who belong to nature, to the Earth and to the universe."[53] Against these thinkers stand notable advocates of animal rights, including, *inter alia*, Jeremy Bentham, Henry Salt, Peter Singer, Tom Regan, and Eugenio Zaffaroni.

Recognizing the ills of anthropocentrism, Judge Tolosa argued that society must transition to an "ecocentric-anthropic worldview ... in which human beings are the main wardens of the universe and the environment, and which promotes a universal and biotic citizenship."[54] Looking to holistically minded environmental ethicists such as Aldo Leopold and Arne Naess,[55] the magistrate asserted that "[w]e are all part of a reconstructive and resilient juridical natural community, and we are citizens that are subjects of proactive rights and members of an organized society that evolves among plants, animals and abiotic agents."[56] This new way of thinking about humans and their place in the world depends on "a sense of respect and responsibility that supersedes the individual and personal level, and that encourages us to see, think and act out of the understanding of the other, of the Earth and of nature."[57] One way to upend the dominant paradigm is to "actively construct ... a conception of the *nature–subject* couplet" (emphasis in original)[58] that will allow us to reconsider our place in the universe and our relationship with nature.

Following his normative assessment of the causes underlying the existential crisis facing humanity, Judge Tolosa then proceeded to explain how survival of the human species depends on a recalibration of relevant legal concepts. In particular, he urged, we need to alter our current conceptions of legal personhood, legal status, and legal rights. This task can be accomplished in three moves. First, we should recognize that nonhuman entities such as corporations and associations are already viewed as legal subjects with legal rights that possess legal personhood despite not being alive or sentient. Second, we must relax the idea that only those entities capable of reciprocating duties are eligible for rights. Third, we can extend legal personhood/status/rights to nature and other beings on the basis of their sentience. This method avoids exclusive reliance on biological, moral, or emotional arguments, and instead advances a "different and creative philosophic-juridical frame."[59]

Judge Tolosa held that domestic and foreign law support the practical implementation of this tripartite strategy. First, Article 655 of the country's Civil Code was amended in 2016 to distinguish entities that exhibit the quality of sentience (i.e., all animals) from those considered merely movable goods (Contreras, 2016, p. 4). The Constitutional Court of Colombia previously found this updated legislation to withstand the test of constitutional scrutiny, explaining that preserving nature is important not only for human survival, but also because parts of nature possess rights of their own.[60] Second, animals are also jointly protected under Articles 8, 79, and 95 of the Colombian Constitution[61] because they are considered a significant part of the state's natural resources. As Judge Tolosa remarked, animals are part of the "natural context, the context where every holder of rights exists and evolves."[62] Third, animal welfare is regulated under Law 84 of 1989,[63] which promulgated the National Animal Protection Statute. Finally, looking to external sources for guidance, Judge Tolosa cited the Ecuadorian, German, and Swiss constitutions, along with the Universal Declaration of Animal Rights (1978), the World Charter for Nature (1982), and a resolution passed by the European Parliament in 1988.

On the basis of the preceding analysis, Judge Tolosa granted a writ of *habeas corpus* to Gómez, who represented Chucho, and ordered the bear's transfer back to the Río Blanco Nature Reserve within 30 days. However, the next month, the Colombian Supreme Court reversed the decision, which Gómez subsequently appealed (Semana, 2017). Then, in early 2020, Chucho's case came to a halt when the Constitutional Court upheld the previous ruling by a 7–2 margin. The president of the Constitutional Court, Judge Gloria Ortiz, stated that animals do not possess a right to freedom because "there is no such category in the Colombian legal system" (City Paper, 2020).

Despite the ultimate failure of Gómez's attempt to free Chucho from captivity, Judge Tolosa's decision illustrates how a multi-faceted animal rights argument might be presented. While the magistrate invoked a properties-based approach through his emphasis on sentience, he also applied a direct/indirect approach, although here the focus was on the survival of all species, not just humans. Judge Tolosa also employed a relational approach when he espoused a decidedly holistic, non-anthropocentric worldview that recognized both the unique status of humans and their concomitant responsibility towards other forms of life. Crucially, he expressed the idea that humans occupy a place among (not above) other entities. Finally, Judge Tolosa adopted a critical legal approach[64] that rejects individualism and Cartesian dualism in favor of radical concepts such as the nature–subject couplet, legal personhood for all sentient beings, and eliminating reciprocal duties from the conditions necessary for an entity to hold rights. Through this multipronged analytical exercise, the magistrate concluded that "[i]n an ethical and ontological sense, rights cannot be an exclusive endowment of human beings."[65]

Although an initial reading of the *Chucho* case suggests that the rights of technological entities are not readily forthcoming, Judge Tolosa's philosophic-juridical approach to animal rights offers a few potential entry points. First, the collapse of the human–environment divide and its replacement with nature–subject couplets leaves open the possibility that other nonhuman beings may emerge as members of the natural juridical community, which includes abiotic agents. Second, although perhaps less likely, were robots to be considered sentient beings, they would be easily accommodated as legal subjects with legal personhood worthy of rights in an ecocentric-anthropic worldview. Third, the removal of reciprocal duties from the list of qualities needed to possess rights would provide an avenue for nonhuman entities to be reclassified as moral patients eligible for moral (and perhaps legal) rights.

A brief comparison of the cases reviewed in this section summarizes how courts in different jurisdictions have decided cases pertaining to animal rights (see Table 3.1).

The above cases provide evidence of both similarity and differentiation in the adjudication of animal rights across legal contexts. First, the causes of action tend to be alleged infringements of human rights applied to nonhuman creatures. This is not surprising, as the petitioners acting on behalf of animals seek to initiate claims that immediately resonate with judges operating within existing legal systems. Second, jurists have looked to a range of sources, especially legal scholarship, to inform their decision-making. However, the degree to which external laws and jurisprudence are referenced varies considerably. Third, virtually all the cases reviewed here appeal in some form or other to properties-based or legal approaches to animal rights, while the relational approach was not embraced with the same fervor. At the same time, properties-based and legal approaches did not yield consistent outcomes, suggesting that their efficacy may depend on the specific context in which they are deployed. As such, their appeal as legal arguments can hardly be described as universal. Critical and pragmatic arguments, on the other hand, while creative, may not be received well in systems less open to new legal ideas. Fourth, a few of the concepts described in Chapter Two proved integral to the decisions reached. These include legal personhood, legal subject, and legal rights. Despite the occasional confusion regarding the definition of a person in the general sense, the evidence suggests that future efforts to secure animal rights should carefully consider how, if at all, these categories apply to nonhumans.

Conclusion

What pearls of wisdom might the theory and practice regarding animal rights contribute to the discussion of rights for technological entities? First, at least some degree of anthropocentrism inevitably invades any moral or legal calculus. None of the approaches discussed above is immune from this charge. However, the extent of anthropocentrism present in the animal rights literature ranges from absolute (i.e., direct/indirect approaches) to marginal (i.e., certain legal approaches). While the act of deciding whether or not an entity should have rights may be an inescapably anthropocentric exercise, we should also recognize the environmental devastation that has befallen the Earth as a result of placing human interests above all others. Therefore, it pays to remain cognizant of anthropocentrism's ongoing destructive capacities, and actively seek to keep them in check so as to avoid further subjugation and eradication of species that inhabit this planet. Applying

Table 3.1 Comparison of animal rights cases

Case	Basis for Legal Action	Source(s) of Legal Reasoning	Animal Rights Approach	Concept(s) Invoked	Decision
Lavery I	Unlawful detention	Domestic case law, legal scholarship	Properties-based	Person, legal personhood, legal rights	Writ of *habeas corpus* denied
Karnail Singh	Violation of law	Domestic law, domestic and foreign case law, religious beliefs, legal scholarship	Properties-based, direct/indirect, legal	Legal personhood, legal rights	All animals granted rights
Sandra I	Unlawful detention	Legal positivism, legal scholarship	Legal	Legal subject, legal rights	Case remanded
Sandra II	Violation of basic rights	Legal positivism, domestic and foreign law	Properties-based, legal	Person, legal subject, legal rights	Sandra recognized as legal subject
Chucho	Unlawful detention	Domestic, foreign, and international law, domestic case law, environmental ethics	Properties-based, direct/indirect, relational, legal	Legal personhood, legal subject, legal rights	Writ of *habeas corpus* granted

this more critical stance to intelligent machines means adopting a weaker, almost enfeebled, version of anthropocentrism similar to the ecocentric-anthropic ethic proposed by Judge Tolosa.[66] Practically speaking, this would require focusing less on qualities other entities might share with humans, and more on ways in which human decision-making could lead to more ethically justifiable determinations, as opposed to continued human supremacy.

Second, the role that duties play in the process of granting rights needs clarification if robots (or any other nonhuman entities for that matter) are ever to enjoy a status above mere machines. On the one hand, duties are the correlative of rights. Watson argues that the capacity to reciprocate duties is necessary for an entity to be considered a moral agent and thus entitled to moral rights. Hohfeld maintains that only human beings can engage in such reciprocal exchanges, limiting the types of beings eligible for legal rights. On the other hand, these logical relationships ignore moral patients, who are capable of holding rights even if they cannot discharge duties, and discount the perspectives of cultures across the world that have afforded legal rights for reasons other than the capacity for mutual accountability. Referring back to the *Chucho* case, Judge Tolosa explicitly entertained the suggestion made by Salt a century earlier to relax the reciprocal duties requirement that has for so long prevented the expansion of legal rights to nonhumans. Once moral patients and legal subjects are no longer shackled by the expectation of reciprocity, radical possibilities, such as rights-bearing technological entities, seem possible.

Third, while some emphasis on the properties of individual entities seems unavoidable, what remains less clear is which qualities should be held in higher regard than others and the relative importance of properties compared to other factors. The answers to these questions have substantial implications for the debate over rights for AI. A promising and increasingly science-based case can be made for promoting sentience above other properties, as suggested by Judge Liberatori's careful consideration of evidence regarding Sandra the orangutan's cognitive abilities. Further, as Boyd illustrates, plenty of research supports the presence of other morally significant properties among members of the animal kingdom. But the properties-based approach still suffers from a general lack of consensus in both philosophy and law, which is due in large part to the yet-unresolved problem of other minds. Even the relational approach, which I have argued may offer the best path forward, relies on the nature of interactions between entities whose responses to each other are shaped by the properties they appear to possess. In addition, the utility of properties-based approaches is further complicated by the argument from marginal cases, which casts doubt on our ability to set an intersubjectively acceptable floor for moral consideration. All of this is to say that the properties quandary remains unsettled in both literature and litigation, although the traits associated with certain entities—be they animal or mechanical—are likely to figure somewhere in moral and legal analyses.

Fourth, the plight of animal rights manifests in the difficulty scholars and jurists experience when attempting to prescribe the boundaries of the moral circle or determine the inclusiveness of legal categories. Philosophers applying properties-based approaches have delimited the kinds of entities worthy of moral rights, but these efforts have relied on subjective, anthropocentric, and empirically problematic individual qualities. Those promoting a relational approach have encouraged a shift away from ontological properties. However, properties have not been completely expelled from the equation. Their implicit relevance has instead introduced a level of analytical uncertainty that might frustrate the practical application of this strategy. Legal scholars have put forth arguments that excel in precision but fall short in terms of inclusiveness. It seems that we can offer greater legal protections for animals found in farms, zoos, or the wild, but not all at once.[67] Courts around the world have also varied considerably as far as coverage is concerned. In *Lavery I*, the imposition of a strict (albeit erroneous) definitional requirement involving the capacity for rights and duties successfully foreclosed the possibility of finding Tommy the chimpanzee to be a legal person entitled to legal rights. In *Karnail Singh*, by contrast, a multi-faceted legal judgment set in the context of developing state granted legal rights to all animals. Thus, both the philosophy and law on animal rights have failed to develop coherent and inclusive programs that could be profitably applied to questions regarding the rights of other entities, such as intelligent machines.

To be direct, the answer to the machine question will not likely come from the intellectual cage of animal rights. However, the animal rights model could prove useful if any of the following three conditions are satisfied. First, if philosophers ever come to an agreement about the existence of properties like consciousness or sentience in animals, perhaps as a result of scientific evidence made possible by new technologies, the inquiry might then shift to whether or not robots possess the same capability. Following Singer, if they are determined to have the same property, there would be no reason why they should not be eligible for moral/psychological and legal personhood, and, consequently, moral or legal rights. Second, if societal needs ever compel moral or legal recognition for AI, this whole debate might prove moot. As discussed in Chapter Two, American jurisprudence supports the idea that corporations and ships should be treated as legal persons for practical reasons. In both *Karnail Singh* and *Chucho*, extending legal rights to animals was similarly deemed necessary, although in these cases it was because of the need to address the impacts of environmental degradation and insure the survival of different species, respectively. Therefore, should society find it useful to consider artificial agents legal persons, the paradigm will adjust accordingly, irrespective of whether or not philosophers have made progress on their own front. Finally, if Western philosophy and its accompanying ontological worldviews ever cede ground to Eastern and critical perspectives, the categorization and treatment of technological beings will be primed for tectonic alteration. Inputs provided by non-Western sources may include recognizing that the human–animal divide is a social construct (Peters, 2016, p. 27), a relic of Cartesian dualism that overlooks temporal and spatial variations in the ontological status of nonhuman creatures; respecting Indigenous knowledge about and ways of conceiving the world; moving away from the focus on human properties towards embracing the kinship that exists among all beings; and replacing individualism with holism. It is this final point that I examine in greater detail in the next chapter, which analyzes developments regarding the rights of nature and how they might contribute to the discussion about rights for artefactual entities.

Notes

1 This quote is often attributed to Henry David Thoreau, but it does not appear in any of his writings. It was likely misattributed to him by Salt (1900), who quoted the passage on page 206 in his article, "The Rights of Animals" (J. Cramer, personal communication, March 16, 2020). Archibald Banks was one of the many pseudonyms used by British journalist Oswald John Frederick Crawfurd.

2 Singer's work is discussed at greater length below.

3 This line of reasoning is known as the "argument from marginal cases" (Narveson, 1977, p. 164). The utility of this argument is discussed in the following section.

4 Although different approaches to environmental ethics, such as biocentrism and ecocentrism, are germane to the discussion of animal rights, for the sake of simplicity I focus mainly on animal ethics in the present chapter. Environmental ethics are pursued in greater detail in the next chapter.

5 See https://www.nonhumanrights.org/.

6 As Wise (2002) explicitly notes, "philosophers argue moral rights; judges decide legal rights. And so I present a legal, and not a philosophical, argument for the dignity-rights of nonhuman animals" (p. 34). As argued in Chapter Two, this intellectual divide has produced much confusion regarding the antecedents of rights and the kinds of entities entitled to them.

7 I elected to include Wise in the section on properties-based approaches to animal rights as opposed to the later section on legal approaches because his argument, while inductively derived from case law, focuses exclusively on how rights emerge from the presence or absence of certain traits.

8 Interestingly, Regan (2004) applies the logic of the negative case to plants, whom, he argues, do not share enough in common with humans to be considered subjects-of-a-life (p. 63).

9 Any semblance of consensus quickly dissipates once both Western and non-Western perspectives on properties are considered on equal footing. For instance, adherents of Vajrayana Buddhism in Bhutan acknowledge a range of sentient beings, including deities that are not visible to the naked eye (Allison, 2019). While this Eastern cosmology assigns value to sentience, Western scholars would likely struggle with the idea of affording moral consideration to entities whose existence cannot be observed directly.

10 See the discussion on the argument from marginal cases earlier in this section.

11 For Watson (1979), the capacity to reciprocate duties is necessary for an entity to possess rights, because only those who act according to the duties they owe others qualify as moral agents, and only moral agents have rights.

12 Tarabout (2019) suggests that compassion might be a source of animal protection that need not be expressed in the language of legal rights. Citing the case of India, he discusses how the constitutional duty to "have compassion for living creatures" (India Const., pt. IVA, art. 51A, §g) has been interpreted by the courts as placing limits on the actions of humans in order to prevent animal suffering.

13 This case is analyzed in detail in the next section.

14 The rights of nature and the issue of legal personhood for natural resources are discussed more extensively in Chapter Four.

15 For Norton (1984), intrinsic value is synonymous with inherent value (p. 137, n11).

16 Legal personhood is extended under the assumption that doing so will advance human interests in the legal domain. The personhood of animals, by contrast, serves to "restrain rather than expand the ways in which human good may be legitimately pursued by means of the instrumental use of animals" (Pietrzykowski, 2017, p. 57).

17 For these authors, the determination of rights is an inquiry separate from the question about an entity's moral agency. However, as demonstrated in Chapter Two, these concepts are logically connected. Duties are correlatives of rights (Hohfeld, 1913) and

both are incidents associated with moral status—agency and patiency. While Marx and Tiefensee focus on whether or not animals or robots qualify as moral agents, nowhere do they mention the possibility that either might be moral patients.

18 Notable exceptions include Maddux (2012) and Staker (2017), who place legal personhood at the center of legal scholarship on animal rights advocacy.

19 Regan, however, does write about moral agency and moral patiency quite extensively. See Sun (2018). Pluhar (1987) maintains that the stronger of Regan's two responses to the argument from marginal cases actually precludes characterizing animals as persons altogether according to Feinberg's (1980) notion of "commonsense" personhood (p. 260), which, ironically, consists of properties similar to those associated with being a subject-of-a-life. This is all to say that perhaps personhood is addressed implicitly in Regan's animal rights philosophy.

20 For instance, I love my dog Shiva, but I cannot be assured that every stranger she meets on our walks will feel similarly towards her.

21 Anthropomorphism in the context of robots is explored in greater detail in Chapter Five.

22 But see Cupp (2017), who argues that "although one could argue that animals have interests and thus should have some form of 'rights' under an expansive view of the interest theory that goes beyond its usual focus on humans and human proxies, such a conclusion is not in any way compelled under the theory" (p. 44).

23 *People ex rel. Nonhuman Rights Project, Inc. v. Lavery* ('Lavery I'), 998 N.Y.S.2d 248 (N.Y. App. Div. 2014), available at https://casetext.com/case/people-v-lavery-6.

24 Ibid., at 249.

25 Ibid.

26 Ibid., at 250.

27 Ibid., at 251.

28 *In re Nonhuman Rights Project, Inc. v. Lavery* ('Lavery II'), 54 N.Y.S.3d 392 (N.Y. App. Div. 2017), available at https://casetext.com/case/nonhuman-rights-project-inc-ex-rel-tommy-v-lavery.

29 Ibid., at 250.

30 Ibid.

31 This case is discussed in greater detail in Chapter Four.

32 *Karnail Singh and others v. state of Haryana* ('Karnail Singh'), Punjab & Haryana High Court, CRR-533-2013 (2019), available at https://www.livelaw.in/pdf_upload/pdf_upload-361239.pdf.

33 See *Animal Welfare Board of India v. Nagaraja and Ors* ('Animal Welfare Board'), 7 SCC 547 (2019), available at https://www.nonhumanrights.org/content/uploads/Animal-Welfare-Board-v-A.-Nagaraja-7.5.2014.pdf.

34 "It shall be the duty of every citizen of India … to protect and improve the natural environment including forests, lakes, rivers and wild life, and to have compassion for living creatures" (India Const., pt. IVA, art. 51A, §g).

35 "The State shall endeavour to organise agriculture and animal husbandry on modern and scientific lines and shall, in particular, take steps for preserving and improving the breeds, and prohibiting the slaughter, of cows and calves and other milch and draught cattle" (India Const., pt. IV, art. 48).

36 "The State shall endeavour to protect and improve the environment and to safeguard the forests and wild life of the country" (India Const., pt. IV, art. 48A).

37 *Karnail Singh*, at 104.

38 *Animal Welfare Board*, at 35. These freedoms are listed in the World Organisation for Animal Health's *Terrestrial Animal Health Code*. See OIE (2019, Chapter 7).

39 Ecocentrism is explained in the following chapter.

40 Citing the Indian Supreme Court's decision in *Animal Welfare Board*, the Court in *Karnail Singh* recounted that the Indian Constitution's fundamental right to property (formerly appearing under Part III, Art. 19(f)) was removed with the adoption of the

44th Amendment in 1978, affording Parliament more flexibility to pass laws pertaining to animal rights. See *Animal Welfare Board*, at 35.

41 *Orangutana Sandra s/ Recurso de Casación s/Habeas Corpus* ('Sandra I'), Cámara Federal de Casación Penal, CCC 68831/2014/CFC1 (2014), available at https://www.nonhumanrights.org/blog/copy-of-argentine-court-ruling/. Translated by Mountain (2014).

42 Ibid., at 86.

43 For instance, news articles featured headlines such as "Court in Argentina Grants Basic Rights to Orangutan" (BBC News, 2014) and "Argentina Grants an Orangutan Human-Like Rights" (Román, 2015).

44 Zaffaroni's scholarship on animal rights has been compared to that of Peter Singer and Tom Regan. See Bandieri (2015).

45 It should be noted that Alejandro Slokar, one of Zaffaroni's co-authors on the book, was also one of the two judges in this case who held that Sandra was entitled to rights.

46 Translation author's own with the assistance of Google Translate.

47 Importantly, an *amparo* is designed to protect all basic rights except for the right to physical freedom, which is addressed through the *habeas corpus* procedure (Adre, 2018, p. 139).

48 *Asociacion de Funcionarios y Abogados por los Derechos de los Animales y Otros contra GBCA sobre amparo* ('Sandra II'), EXPTE. A2174-2015/0 (2015), available at https://ijudicial.gob.ar/wp-content/uploads/2015/10/Sentencia-Orangutana.pdf.

49 Liberatori explicitly borrowed this phrase from Italian lawyer and sociologist Valerio Pocar. See Pocar (2013).

50 For a more detailed analysis of how changes in the French Civil Code affect the legal status of animals, see Neumann (2015).

51 Ibid., at 10. Translation author's own with the assistance of Google Translate.

52 AHC4806–2017 ('Chucho'), Radicación no. 17001–22–13–000–2017–00468–02, available at http://static.iris.net.co/semana/upload/documents/radicado-n-17001-22-13-000-2017-00468-02.pdf. Translated by Javier Salcedo, available at https://www.nonhumanrights.org/content/uploads/Translation-Chucho-Decision-Translation-Javier-Salcedo.pdf.

53 Ibid., at 4.

54 Ibid., at 5.

55 These and other environmental ethicists are discussed more extensively in the next chapter.

56 *Chucho*, at 5.

57 Ibid., at 6.

58 Ibid.

59 Ibid., at 8.

60 Corte Constitucional, Sentencia C-041 (2017), available at https://www.animallaw.info/case/sentencia-c-041-2017.

61 "It is the obligation of the State and of individuals to protect the cultural and natural assets of the nation" (Colombia Const., tit. I, art. 8); "Every individual has the right to enjoy a healthy environment. An Act shall guarantee the community's participation in the decisions that may affect it. It is the duty of the State to protect the diversity and integrity of the environment, to conserve the areas of special ecological importance, and to foster education for the achievement of these ends" (Colombia Const., tit. II, ch. III, art. 79); "The following are duties of the individual and of the citizen: ... To protect the country's cultural and natural resources and to keep watch that a healthy environment is being preserved (Colombia Const., tit. II, ch. V, art. 95, §8).

62 *Chucho*, at 9.

63 Law 84 of 1989, Official Diary 39120, of December 27, 1989, available at https://www.alcaldiabogota.gov.co/sisjur/normas/Norma1.jsp?i=8242.

64 Critical environmental law is described in the following chapter.

65 *Chucho*, at 13.

66 I write "weaker" and not "weak" so as to distinguish my argument from that of Epting (2010), who writes that weak anthropocentrism dictates "approach[ing] technology in a manner that is conducive to the permanence of genuine human life, which requires ecological sustainability" (p. 21).

67 The lone exception worth mentioning here is the approach advocated by Abate and Crowe, although their argument rests in part on making an analogy between animals and humans by virtue of their shared properties (i.e., sentience). This argument remains suspect for reasons specified above.

References

Aaltola, E. (2008). Personhood and Animals. *Environmental Ethics*, *30*(2), 175–193.

Abate, R. S., & Crowe, J. (2017). From Inside the Cage to Outside the Box: Natural Resources as a Platform for Nonhuman Animal Personhood in the U.S. and Australia. *Global Journal of Animal Law*, *5*(1), 54–78.

Abramowicz, M., & Stearns, M. (2005). Defining Dicta. *Stanford Law Review*, *57*(4), 953–1094.

Adre, G. R. (2018). El Amparo en la Justicia Argentina. ¿La Vía Idónea para el Reconocimiento de los Derechos de los ANH? *Derecho Animal: Forum of Animal Law Studies*, *9*(4), 138–150.

Allison, E. (2019). Deity Citadels: Sacred Sites of Bio-Cultural Resistance and Resilience in Bhutan. *Religions*, *10*(4), 268.

Animal Legal Defense Fund (2020). *2019 U.S. Animal Protection Laws Ranking Report*. Retrieved from https://aldf.org/wp-content/uploads/2020/02/2019-Animal-Protection-US-State-Laws-Rankings-Report.pdf.

Bandieri, L. M. (2015). Los animales, ¿Tienen Derechos? *Prudentia Iuris*, *79*, 33–56.

Banks, A. (1874). Birds and Beasts in Captivity. *The New Quarterly Magazine*, *2*, 793–819.

Bao-Er (2014). China's Confucian Horses: The Place of Nonhuman Animals in a Confucian World Order. In N. Dalal & C. Taylor (Eds.), *Asian Perspectives in Animal Ethics: Rethinking the Nonhuman* (pp. 73–92). Routledge.

BBC News (2014, December 21). Court in Argentina Grants Basic Rights to Orangutan. *BBC News*. Retrieved from https://www.bbc.com/news/world-latin-america-30571577.

Bentham, J. (1879). *An Introduction to the Principles of Morals and Legislation*. Clarendon Press.

Borkfelt, S. (2011). What's in a Name?—Consequences of Naming Non-Human Animals. *Animals*, *1*(1), 116–125.

Boyd, D. R. (2017). *The Rights of Nature: A Legal Revolution That Could Save the World*. ECW.

Boyer, P. (1996). What Makes Anthropomorphism Natural: Intuitive Ontology and Cultural Representations. *The Journal of the Royal Anthropological Institute*, *2*(1), 83–97.

Bradshaw, K. (2018). Animal Property Rights. *University of Colorado Law Review*, *89*(3), 809–862.

Bryant, T. L. (2008). Sacrificing the Sacrifice of Animals: Legal Personhood for Animals, the Status of Animals as Property, and the Presumed Primacy of Humans. *Rutgers Law Journal*, *39*(2), 247–330.

Callicot, J. B. (1995). Intrinsic Values in Nature: A Metaethical Analysis. *The Electronic Journal of Analytical Philosophy*, *3*(5). Retrieved from https://ejap.louisiana.edu/EJAP/1995.spring/callicott.1995.spring.

Calverley, D. J. (2006). Android Science and Animal Rights: Does an Analogy Exist? *Connection Science*, *18*(4), 403–417.

Chan, K. M. A. (2011). Ethical Extensionism under Uncertainty of Sentience: Duties to Non-Human Organisms without Drawing a Line. *Environmental Values*, *20*(3), 323–346.

Chiew, F. (2014). Posthuman Ethics with Cary Wolfe and Karen Barad: Animal Compassion as Trans-Species Entanglement. *Theory, Culture and Society*, *31*(4), 51–69.

Choplin, L. (2019, August 19). NhRP Addresses Highest Court in Colombia in Chucho Bear Rights Case. *Nonhuman Rights Project*. Retrieved from https://www.nonhuman rights.org/blog/chucho-supreme-court-hearing-colombia/.

City Paper (2020, January 23). Colombia's Constitutional Court Denies Habeas Corpus for Andean Bear. *The City Paper Bogotá*. Retrieved from https://thecitypaperbogota.com/news/colombias-constitutional-court-denies-habeas-corpus-for-andean-bear/23781.

Coeckelbergh, M., & Gunkel, D. J. (2014). Facing Animals: A Relational, Other-Oriented Approach to Moral Standing. *Journal of Agricultural and Environmental Ethics*, *27*(5), 715–733.

Colombia Const., tit. I, art. 8.

Colombia Const., tit. II, ch. III, art. 79.

Colombia Const., tit. II, ch. V, art. 95, §8.

Contreras, C. (2016). Sentient Beings Protected by Law. Analysis of Recent Changes in Colombian Animal Welfare Legislation. *Global Journal of Animal Law*, *2*, 1–19.

Cottingham, J. (1978). "A Brute to the Brutes?" Descartes' Treatment of Animals. *Philosophy*, *53*(206), 551–559.

Crowe, J. (2008). Levinasian Ethics and Animal Rights. *Windsor Yearbook of Access to Justice*, *26*(2), 313–328.

Cullinan, C. (2003). *Wild Law: A Manifesto for Earth Justice*. Green Books.

Cupp, R. L. (2017). Cognitively Impaired Humans, Intelligent Animals, and Legal Personhood. *Florida Law Review*, *69*(2), 465–518.

de Baggis, G. F. (2015). Solicitud de Hábeas Corpus para la Orangután Sandra: Comentario a propósito de la Sentencia de la Cámara Federal de Casación Penal de la Ciudad Autónoma de Buenos Aires, de 18 de diciembre de 2014. *Derecho Animal: Forum of Animal Law Studies*, *6*(1), 1–8.

DeGrazia, D. (2002). *Animal Rights: A Very Short Introduction*. Oxford University Press.

Dennett, D. C. (1981). *Brainstorms: Philosophical Essays of Mind and Psychology* (1st ed.). MIT Press.

Derrida, J. (2002). The Animal That Therefore I Am (More to Follow) (D. Wills, Trans.). *Critical Inquiry*, *28*(2), 369–418.

Descartes, R. (1924). *Discourse on Method* (J. Veitch, Trans.). Open Court. (Original work published 1637).

Descartes, R. (1970). *Descartes: Philosophical Letters* (A. Kenny, Trans.). Clarendon Press.

Dombrowski, D. A. (1997). *Babies and Beasts: The Argument from Marginal Cases*. University of Illinois Press.

Doshi-Velez, F., & Kortz, M. (2017). *Accountability of AI Under the Law: The Role of Explanation*. Berkman Klein Center for Internet & Society. Harvard University. Retrieved from https://dash.harvard.edu/bitstream/handle/1/34372584/2017-11_aiex plainability-1.pdf.

Elassar, A. (2019, November 9). Sandra the Orangutan, Freed from a Zoo After Being Granted "Personhood," Settles into Her New Home. *CNN*. Retrieved from https://www.cnn.com/2019/11/09/world/sandra-orangutan-florida-home-trnd/index.html.

Epting, S. (2010). Questioning Technology's Role in Environmental Ethics: Weak Anthropocentrism Revisited. *Interdisciplinary Environmental Review*, *11*(1), 18–26.

European Parliament (2017). *Report with Recommendations to the Commission on Civil Law Rules on Robotics* (No. A8-0005/2017). Retrieved from https://www.europarl.euro pa.eu/doceo/document/A-8-2017-0005_EN.pdf.

Feinberg, J. (1980). Abortion. In T. Regan (Ed.), *Matters of Life and Death* (2nd ed., pp. 256–293). Random House.

Feinberg, J. (2013). The Rights of Animals and Unborn Generations. In R. Shafer-Landau (Ed.), *Ethical Theory: An Anthology* (2nd ed., pp. 372–380). Wiley-Blackwell.

Framarin, C. (2014). Atman, Identity, and Emanation: Arguments for a Hindu Environmental Ethic. In J. B. Callicot & J. McRae (Eds.), *Environmental Philosophy in Asian Traditions of Thought* (pp. 25–52). State University of New York Press.

Frey, R. G. (1980). *Interests and Rights: The Case Against Animals*. Clarendon Press.

Frohlich, X. (2016). The Informational Turn in Food Politics: The Us FDA's Nutrition Label as Information Infrastructure. *Social Studies of Science*, *47*(2), 145–171.

GAP Project (2017, July 18). Argentine Judge Refuses to Transfer Orangutan Sandra to Great Apes Sanctuary of Sorocaba, Brazil. *GAP Project*. Retrieved from https://www .projetogap.org.br/en/noticia/argentine-judge-refuses-to-transfer-orangutan-sandra-to -great-apes-sanctuary-of-sorocaba-brazil/.

Girgen, J. (2003). The Historical and Contemporary Prosecution and Punishment of Animals. *Animal Law*, *9*, 97–133.

Goetz, J. L., Keltner, D., & Simon-Thomas, E. (2010). Compassion: An Evolutionary Analysis and Empirical Review. *Psychological Bulletin*, *136*(3), 351–374.

Grear, A. (2013). Law's Entities: Complexity, Plasticity and Justice. *Jurisprudence*, *4*(1), 76–101.

Gunkel, D. J. (2012). *The Machine Question: Critical Perspectives on AI, Robots, and Ethics*. MIT Press.

Hajjar Leib, L. (2011). *Human Rights and the Environment: Philosophical, Theoretical and Legal Perspectives*. Martinus Nijhoff.

Hallevy, G. (2010). "I, Robot-I, Criminal"—When Science Fiction Becomes Reality: Legal Liability of AI Robots Committing Criminal Offenses. *Syracuse Science and Technology Law Reporter*, *22*, 1–37.

Harrison, F. (1904). Duties of Man to the Lower Animals. *The Humane Review*, *5*, 1–10.

Harrison, P. (1992). Descartes on Animals. *The Philosophical Quarterly*, *42*(167), 219–227.

Hawkins, R. D., Williams, J. M., & Scottish SPCA. (2017). Childhood Attachment to Pets: Associations between Pet Attachment, Attitudes to Animals, Compassion, and Humane Behaviour. *International Journal of Environmental Research and Public Health*, *14*(5), 490.

Hervy, E. (2015). France's Glavany Amendment: Animals Now Considered "Living Beings Gifted Sentience". *The Speaker News Journal*. Retrieved from https://thespea kernewsjournal.com/headlines/frances-glavany-amendment-animals-now-considered -living-beings-gifted-sentience/.

Hogan, K. (2017). Is the Machine Question the Same Question as the Animal Question? *Ethics and Information Technology*, *19*(1), 29–38.

Hohfeld, W. N. (1913). Some Fundamental Conceptions as Applied in Judicial Reasoning. *Yale Law Journal*, *23*(1), 16–59.

India Const., pt. IV, art. 48.

India Const., pt. IV, art. 48A.

India Const., pt. IVA, art. 51A, §g.

Kant, I. (1963). *Lectures on Ethics* (L. Infield, Trans.). Harper & Row.

Kant, I. (1997). *Lectures on Metaphysics* (K. Ameriks & S. Naragon, Trans.). Cambridge University Press.

Kelch, T. G. (1999). The Role of the Rational and the Emotive in a Theory of Animal Rights. *Boston College Environmental Affairs Law Review*, *27*(1), 1–41.

Kniess, J. (2019). Bentham on Animal Welfare. *British Journal for the History of Philosophy*, *27*(3), 556–572.

Lehmann, J., Breuker, J., & Brouwer, B. (2003). *Causation in AI & Law*, 1–34.

Leslie, J., & Sunstein, C. R. (2007). Animal Rights without Controversy. *Law and Contemporary Problems*, *70*(1), 117–138.

Linzey, A. (1989). The Theos-Rights of Animals. In T. Regan & P. Singer (Eds.), *Animal Rights and Human Obligations* (2nd ed., pp. 134–138). Prentice Hall.

Linzey, A., & Regan, T. (1990). Introduction to Part One. In A. Linzey & T. Regan (Eds.), *Animals and Christianity: A Book of Readings* (pp. 3–5). Wipf & Stock.

Maddux, E. A. (2012). Time to Stand: Exploring the Past, Present, and Future of Nonhuman Animal Standing. *Wake Forest Law Review*, *47*(5), 1243–1267.

Malik, S. (2019, June 1). Animals Are 'Legal Persons', All Citizens Their Guardians, Says HC. *The Tribune*. Retrieved from https://www.tribuneindia.com/news/archive/animals-are-legal-persons-all-citizens-their-guardians-says-hc-781738.

Marx, J., & Tiefensee, C. (2015). Of Animals, Robots and Men. *Historical Social Research / Historische Sozialforschung*, *40*(4), 70–91.

Mawdsley, E. (2006). Hindu Nationalism, Neo-Traditionalism and Environmental Discourses in India. *Geoforum*, *37*(3), 380–390.

McGrath, J. F. (2011). Robots, Rights, and Religion. In J. F. McGrath (Ed.), *Religion and Science Fiction* (pp. 118–153). Pickwick.

Michalczak, R. (2017). Animals' Race Against the Machines. In V. A. J. Kurki & T. Pietrzykowski (Eds.), *Legal Personhood: Animals, Artificial Intelligence and the Unborn* (pp. 91–101). Springer.

Morito, B. (2003). Intrinsic Value: A Modern Albatross for the Ecological Approach. *Environmental Values*, *12*(3), 317–336.

Mountain, M. (2014, December 23). Translation of Argentine Court Ruling. *Nonhuman Rights Blog*. Retrieved from https://www.nonhumanrights.org/blog/copy-of-argentine-court-ruling/.

Nagel, T. (1974). What Is It Like to Be a Bat? *The Philosophical Review*, *83*(4), 435–450.

Narveson, J. (1977). Animal Rights. *Canadian Journal of Philosophy*, *7*(1), 161–178.

Neumann, J.-M. (2015). The Legal Status of Animals in the French Civil Code: The Recognition by the French Civil Code That Animals Are Living and Sentient Beings: Symbolic Move, Evolution or Revolution? *Global Journal of Animal Law*, *1*, 1–13.

Nonhuman Rights Project (n.d.). *Client, Tommy (Chimpanzee): The NhRP's First Client*. Nonhuman Rights Project. Retrieved March 7, 2020, from https://www.nonhumanrights.org/client-tommy/.

Norton, B. G. (1984). Environmental Ethics and Weak Anthropocentrism. *Environmental Ethics*, *6*(2), 131–148.

O'Donnell, E. L. (2018). At the Intersection of the Sacred and the Legal: Rights for Nature in Uttarakhand, India. *Journal of Environmental Law*, *30*(1), 135–144.

OIE (World Organisation for Animal Health) (2019). *Terrestrial Animal Health Code: Vols. I and II* (28th ed.). Retrieved from https://www.oie.int/en/standard-setting/terrestrial-code/access-online/.

Park, Y. S., & Valentino, B. (2019). Animals Are People Too: Explaining Variation in Respect for Animal Rights. *Human Rights Quarterly, 41*(1), 39–65.

Peters, A. (2016). Liberté, Égalité, Animalité: Human–Animal Comparisons in Law. *Transnational Environmental Law, 5*(1), 25–53.

Peters, A. (2018). Rights of Human and Nonhuman Animals: Complementing the Universal Declaration of Human Rights. *AJIL Unbound, 112*, 355–360.

Peterson, A. L. (2013). *Being Animal: Beasts and Boundaries in Nature Ethics*. Columbia University Press.

Phelps, N. (2002). *The Dominion of Love: Animal Rights According to the Bible*. Lantern Books.

Phelps, N. (2004). *The Great Compassion: Buddhism and Animal Rights*. Lantern Books.

Pietrzykowski, T. (2017). The Idea of Non-Personal Subjects of Law. In V. A. J. Kurki & T. Pietrzykowski (Eds.), *Legal Personhood: Animals, Artificial Intelligence and the Unborn* (pp. 49–67). Springer.

Pluhar, E. B. (1987). The Personhood View and the Argument from Marginal Cases. *Philosophica, 39*(1), 23–38.

Pocar, V. (2013). *Los Animales no Humanos: Por una Sociología de los Derechos* (L. N. Lora, Trans.). Ad-Hoc.

Regan, T. (1979). An Examination and Defense of One Argument Concerning Animal Rights. *Inquiry: An Interdisciplinary Journal of Philosophy, 22*(1–4), 189–219.

Regan, T. (1983). *The Case for Animal Rights*. University of California Press.

Regan, T. (1987). The Case for Animal Rights. In M. W. Fox & L. D. Mickley (Eds.), *Advances in Animal Welfare Science 1986/87* (pp. 179–189). Martinus Nijhoff Publishers.

Regan, T. (1990). Christianity and Animal Rights: The Challenge and Promise. In C. Birch, W. Eaken, & J. B. McDaniel (Eds.), *Liberating Life: Contemporary Approaches in Ecological Theology* (pp. 73–87). Orbis Books.

Regan, T. (2004). *Empty Cages: Facing the Challenge of Animal Rights*. Rowman & Littlefield.

Román, V. (2015, January 9). Argentina Grants an Orangutan Human-Like Rights. *Scientific American*. Retrieved from https://www.scientificamerican.com/article/ar gentina-grants-an-orangutan-human-like-rights/.

Rowlands, M. (1997). Contractarianism and Animal Rights. *Journal of Applied Philosophy, 14*(3), 235–247.

Salt, H. S. (1900). The Rights of Animals. *International Journal of Ethics, 10*(2), 206–222.

Schauer, F. (2000). *The Politics and Incentives of Legal Transplantation* (CID Working Paper Series 2000.44). Harvard University. Retrieved from http://nrs.harvard.edu/urn-3 :HUL.InstRepos:39526299.

Semana (2017, August 18). El Oso de la Justicia: Chucho se Queda en el Zoológico de Barranquilla. *Semana*. Retrieved from https://www.semana.com/nacion/articulo/niega-corte-suprema-habeas-corpus-al-oso-de-anteojos-chucho/536929.

Singer, P. (1974). All Animals Are Equal. *Philosophic Exchange, 5*(1), 103–116.

Singer, P. (1979). *Practical Ethics*. Cambridge University Press.

Singer, P., & Sagan, A. (2009, December 14). When Robots Have Feelings. *The Guardian*. Retrieved from https://www.theguardian.com/commentisfree/2009/dec/14/rage-agains t-machines-robots.

Slaughter, A.-M. (1994). A Typology of Transjudicial Communication. *University of Richmond Law Review, 29*, 99–138.

Snyder, B. F. (2017). The Darwinian Nihilist Critique of Environmental Ethics. *Ethics and the Environment*, *22*(2), 59–78.

Staker, A. (2017). Should Chimpanzees Have Standing? The Case for Pursuing Legal Personhood for Non-Human Animals. *Transnational Environmental Law*, *6*(3), 485–507.

Stein, E. (2017, March 27). *Letter to First Department Re: Tommy and Kiko*. Retrieved from https://www.nonhumanrights.org/content/uploads/Letter-to-First-Dept-re-Tommy-and-Kiko-3.27.17-FINAL-1.pdf.

Stucki, S., & Herrera, J. C. (2017, November 3). Habea(r)s Corpus: Some Thoughts on the Role of Habeas Corpus in the Evolution of Animal Rights. *International Journal of Constitutional Law Blog*. Retrieved from http://www.iconnectblog.com/2017/11/habears-corpus-some-thoughts-on-the-role-of-habeas-corpus-in-the-evolution-of-animal-rights/.

Sun, Y. (2018). The Edge of "Animal Rights." *Journal of Agricultural and Environmental Ethics*, *31*(5), 543–557.

Sunstein, C. R. (2000). Standing for Animals (with Notes on Animal Rights): A Tribute to Kenneth L. Karst. *UCLA Law Review*, *47*, 1333–1368.

Szűcs, E., Geers, R., Jezierski, T., Sossidou, E. N., & Broom, D. M. (2012). Animal Welfare in Different Human Cultures, Traditions and Religious Faiths. *Asian-Australasian Journal of Animal Sciences*, *25*(11), 1499–1506.

Tanner, J. K. (2009). The Argument from Marginal Cases and the Slippery Slope Objection. *Environmental Values*, *18*(1), 51–66.

Tarabout, G. (2019). Compassion for Living Creatures in Indian Law Courts. *Religions*, *10*(6), 383.

Thompson, S. (2019). Supporting Ape Rights: Finding the Right Fit Between Science and the Law. *ASEBL Journal*, *14*(1), 3–24.

Torrance, S. (2013). Artificial Agents and the Expanding Ethical Circle. *AI and Society*, *28*(4), 399–414.

Tymowski, G. (2013). The Virtue of Compassion: Animals in Sport, Hunting as Sport, and Entertainment. In J. Gillett & M. Gilbert (Eds.), *Sport, Animals, and Society* (pp. 140–154). Routledge.

Universal Declaration of Animal Rights (proposed Oct. 15, 1978).

Vallely, A. (2014). Being Sentiently with Others: The Shared Existential Trajectory Among Humans and Nonhumans in Jainism. In N. Dalal & C. Taylor (Eds.), *Asian Perspectives on Animal Ethics: Rethinking the Nonhuman* (pp. 38–55). Routledge.

Vayr, B. (2017). Of Chimps and Men: Animal Welfare vs. Animal Rights and How Losing the Legal Battle May Win the Political War for Endangered Species Notes. *University of Illinois Law Review*, *2*, 817–876.

Vogel, D. (2001). How Green Is Judaism? Exploring Jewish Environmental Ethics. *Business Ethics Quarterly*, *11*(2), 349–363.

Waldau, P. (2000). Buddhism and Animal Rights. In D. Keown (Ed.), *Contemporary Buddhist Ethics* (pp. 81–112). Curzon Press.

Watson, R. A. (1979). Self-Consciousness and the Rights of Nonhuman Animals and Nature. *Environmental Ethics*, *1*(2), 99–129.

Weston, A. (1985). Beyond Intrinsic Value: Pragmatism in Environmental Ethics. *Environmental Ethics*, *7*(4), 321–339.

Wilson, H. L. (2017). The Green Kant: Kant's Treatment of Animals. In P. Pojman & K. McShane (Eds.), *Food Ethics* (2nd ed., pp. 5–13). Cengage Learning.

Wise, S. M. (2002). *Drawing the Line: Science and the Case for Animal Rights*. Perseus Books.

Wise, S. M. (2013). Nonhuman Rights to Personhood. *Pace Environmental Law Review*, *30*(3), 1278–1290.

Wise, S. M. (2015a, January 12). Sandra: The Plot Thickens. *Nonhuman Rights Blog*. Retrieved from https://www.nonhumanrights.org/blog/sandra-the-plot-thickens/.

Wise, S. M. (2015b, March 6). Update on the Sandra Orangutan Case in Argentina. *Nonhuman Rights Blog*. Retrieved from https://www.nonhumanrights.org/blog/update -on-the-sandra-orangutan-case-in-argentina/.

World Charter for Nature, GA res. 37/7, 37 UN GAOR, Supp. (No. 51) at 17, UN Doc. A/37/51 (1982).

Wu, S.-C. (2014). Anthropocentric Obsession: The Perfuming Effects of Vāsanā (Habit-Energy) in Ālayavijñāna in the Lan 'Kāvatāra Sūtra. *Contemporary Buddhism*, *15*(2), 416–431.

Zaffaroni, E. R. (2011). *La Pachamama y El Humano*. Ediciones Madres de Plaza de Mayo.

Zaffaroni, E. R., Alagia, A., & Slokar, A. (2002). *Derecho Penal: Parte General* (2nd ed.). Ediar.

Zamir, T. (2007). *Ethics and the Beast: A Speciesist Argument for Animal Liberation*. Princeton University Press.

4 The rights of nature

Ethics, law, and the Anthropocene

Mitákuye oyás'į
(We are all related)

(Lakota people of North America)

The rights of nature (RoN) movement, with its origins in Indigenous traditional knowledge and ancestral cultures, has obtained concrete expression in courts, constitutions, and citizen referenda in numerous places around the world. The ontological premise underlying this global initiative is that the Cartesian separation between man and nature is illusory; all organic life is intimately connected. At the same time, the arrival of the Anthropocene has exposed frailties in the concept of legal personhood and invited a debate over the boundaries of nature itself. Responding to these developments, in this chapter I detail how the RoN movement, scholarship in environmental ethics and law, and recent case law expand the scope of rights to include nonhuman entities. Beginning with an overview of the RoN movement, the chapter proceeds by reviewing biocentric and ecocentric environmental ethics and the extent to which both offer space for extending rights to nonhuman entities. Next, relying on advancements in critical environmental legal scholarship, writing on law in the Anthropocene, and New Materialism, I argue for an expansive definition of the environment and observe that the collapse of the human/nonhuman binary opens up the possibility of widening the range of entities that qualify for rights. Then, I examine how rights have already been extended to natural nonhuman entities under the auspices of the RoN, which have been adjudicated successfully in courts within Ecuador, Colombia, and India. Finally, from the foregoing evidence I demonstrate how a critical, Anthropocene-informed approach to environmental law supports widening the concept of rights to include artefactual entities that exist in and comprise the larger built environment, such as robots.

Origins of the rights of nature

The notion that nature possesses rights emerged from the uncommon and timely union of non-Western and Western ideas (Kauffman & Martin, 2017).

This marriage occurs at a time of acute environmental crisis on a global scale. Indigenous cultures and ancient religions have long observed a complex relationship between humans and the environment involving respect for all forms of life commensurate with human responsibilities towards nature (Boyd, 2017, p. xxix). For instance, the concept of *Buen Vivir* (living well) that emerged in Latin America as a reaction to the negative consequences of development was imbued with Indigenous meaning through the terms *sumak kawsay* (kichwa) and *suma qamaña* (aymara), which emphasize a holistic view of the world based on reciprocity between humans and nature (Gudynas, 2011, p. 442). These terms have influenced the content of and underlying inspiration for the current Ecuadorian Constitution[1] and recent Bolivian statutes,[2] respectively (Calzadilla & Kotzé, 2018, p. 399). Yet for hundreds of years, Indigenous knowledge has been subjugated by powerful Western interests (Borrows, 1997, p. 425). Its newfound exposure and resonance across the world augurs the arrival of a mounting popular resistance to widespread ecological destruction that decries the impotence of current environmental law to provide adequate protection for nature (Collins, 2019).

Numerous examples of the conflict between Indigenous cultures and modern developmental imperatives abound. In the United States, thousands joined the Standing Rock Sioux Tribe for several months in 2016 to protest construction of the Dakota Access pipeline over Native burial sites (Meyer, 2016). In Ecuador, Indigenous communities marched together to oppose environmentally destructive mining activities on Native territory (Brown, 2018). In Brazil, the Karipuna people witnessed deforestation due to logging and raging wildfires in their sections of the Amazonian rainforest (Magalhaes & Pearson, 2019). In New Zealand, a massive housing development project situated on sacred Māori land drew thousands of protestors (Reuters, 2019). In these situations and others, environmental degradation caused by economic development has posed an existential threat to Indigenous peoples, igniting concerns that violations of their traditional relationships with nature may amount to "cultural genocide" (Kingston, 2015, p. 63).

The protection of Native communities and the environments in which they live is often frustrated by the fact that much Indigenous knowledge is not written, but rather transferred orally (Maurial, 1999, p. 63). As such, Indigenous ideas often lack the kind of concrete instantiation or legalization that grants them legitimacy in the Western legal order. One way that Indigenous communities have found success in gaining recognition for their rights is through strategic framing. By explicitly articulating how provisions in international human rights treaties relate to the specific concerns of Native peoples, they have been able to marshal support for legal protections at the international level (Morgan, 2004). Another way that Indigenous knowledge has penetrated legal discourse is through its deliberate incorporation into Western institutions by advocates and allies. For example, Ecuador granted rights to nature in its 2008 constitution due in large part to the opportunistic combination of "radical Western ecological perspectives, politicized indigenous beliefs, and legal rights discourse" advanced by activists and lawyers operating within the dominant legal framework (Akchurin, 2015, p. 961).

Recently in the United States, Native groups have applied traditional concepts to modern environmental problems by legalizing the rights of natural entities. For example, in 2016 the Ho-Chunk Nation of Wisconsin included the RoN in their tribal constitution in order to counteract oil and gas exploration on sacred lands (Margil, 2016). The next year, in an effort to call attention to and prevent further hydraulic fracturing (or "fracking") on or around their reservation, the Ponca Nation of Oklahoma passed a statute recognizing the RoN (Biggs, 2017). At the end of 2018, the White Earth band of Ojibwe passed a law giving rights to *manoomin* (or "wild rice") to protect this traditional dietary staple from industrial activities that harm water quality and introduce genetically modified organisms into the ecosystem (LaDuke, 2019). In each of these instances, Indigenous peoples pushed back against exploitation and degradation of their resources by translating cultural concepts into tools intelligible to modern legal systems—rights.

On the Western front, perhaps no intellectual input has proven more influential in the RoN movement than Christopher Stone's (1972) seminal article "Should Trees Have Standing?—Toward Legal Rights for Natural Objects."[3] In this provocative work, Stone proposed to grant legal rights not only to natural objects, but "to the natural environment as a whole" (1972, p. 456). Extending rights to nature, Stone argued, requires creative interpretation of both legal-operational and psychic or socio-psychic concerns. The former entails determining whether or not an entity qualifies as a holder of legal rights, a project quite similar to deciphering whether or not an entity satisfies the criteria for legal personhood. According to Stone, an entity may enjoy legal rights when 1) an authoritative body is willing to review actions that threaten it; 2) it can institute legal actions on its own accord (i.e., judicial standing); 3) a court considers injury to the entity when granting legal relief; and 4) the relief provided by the court benefits the entity (1972, p. 458). The latter involves "effecting a radical shift in our feelings about 'our' place in the rest of Nature" (Stone, 1972, p. 495). This shift would advance a kind of ecological consciousness whereby the Earth is viewed as a single organism and humans are perceived as different from, but functionally part of, nature.

That Stone's innovative and thought-provoking article found its way into American jurisprudence was no accident. The staff of the *University of Southern California Law Review* deliberately inserted the piece into a special issue of the journal, which was reviewed by Supreme Court Justice William O. Douglas (Stone, 1987, p. 4). Justice Douglas, an avid environmentalist, thus read an early version of the essay prior to the Supreme Court rendering its judgment in *Sierra Club v. Morton*,[4] a case regarding an organization's standing to sue on the basis that its members would suffer an injury in the event that a development project in Mineral King Valley would proceed as planned. Although the Court ultimately decided that Sierra Club did not possess standing due to the group's general and not particularized injury, Justice Douglas penned a dissent that would become legendary in modern environmental law because of its assertion that nature itself might qualify as a legal subject:

Contemporary public concern for protecting nature's ecological equilibrium should lead to the conferral of standing upon environmental objects to sue for their own preservation. See Stone, Should Trees Have Standing?—Toward Legal Rights for Natural Objects, 45 S. Cal. L. Rev. 450 (1972). This suit would therefore be more properly labeled as *Mineral King* v. *Morton.*[5]

Justice Douglas went on to explain the precedent that exists for granting nature legal personhood, citing the examples of ships and corporations that have enjoyed such elevated status when deemed convenient for adjudicatory purposes, and arguing in favor of extending the same recognition to natural entities like rivers. In particular, Justice Douglas reasoned that a river

> speaks for the ecological unit of life that is part of it. Those people who have a meaningful relation to that body of water … must be able to speak for the values which the river represents and which are threatened with destruction.[6]

Importantly, in the preceding statement, the famed jurist observed a kind of reflexive relationship whereby, ontologically speaking, nature encompasses all life that comprise it while, practically speaking, those with direct ties to that part of nature must be eligible to defend its interests in human institutions. Here, Douglas afforded Stone's legal-operational and socio-psychic concerns concrete, jurisprudential instantiation.

Stone's influential article and Douglas's practical application of the RoN in a U.S. Supreme Court decision paved the way for further uptake of the concept in legal systems around the world. In 2006, Tamaqua Borough in Pennsylvania, U.S., became the first municipality to adopt a RoN ordinance (Cano-Pecharroman, 2018, p. 4). The ordinance, which sought to enjoin the dumping of sewage sludge by corporations, includes in the long form of its title a legislative intent motivated by the RoN: "By Recognizing and Enforcing the Rights of Residents to Defend Natural Communities and Ecosystems" (Tamaqua Borough, 2006, p. 1). Later in the ordinance, the scope of entities granted legal personhood is explicitly defined: "Borough residents, natural communities, and ecosystems shall be considered to be 'persons' for purposes of the enforcement of the civil rights of those residents, natural communities, and ecosystems" (Tamaqua Borough, 2006, p. 6).

Instrumental in this initial effort was the training and advocacy provided by the Community Environmental Legal Defense Fund (CELDF), a U.S.-based public interest law firm.[7] CELDF has since assisted dozens of local communities within the United States, as well as other countries, with writing and garnering support for RoN laws. Only two years after the Tamaqua Borough ordinance passed, Ecuador enacted RoN in its 2008 constitution with the help of CELDF. In this capacity, CELDF consulted with Indigenous members of the Ecuadorian Constitutional Assembly and drafted the national charter's RoN provisions (O'Gorman, 2017, p. 447).

While both Indigenous ideas and Western legal thought have played a major role in the inception and spread of RoN initiatives, an academic debate among environmental ethicists regarding human duties towards the environment and the boundaries of the moral circle has quietly occurred in parallel ever since the 1960s. In the next section, I review the main arguments put forth by some of the most notable participants in this ongoing discussion in order to determine the range of entities potentially eligible for consideration within their ethical universes and the kind of rights that might be bestowed upon them.

Environmental ethics: biocentrism, ecocentrism, and nonhuman rights

The field of modern environmental ethics has engaged in a robust debate over the ideal ethical orientation towards nature and its inhabitants at least since Aldo Leopold (1966) encouraged "thinking like a mountain" (p. 137). An anthropocentric ethic places human interests and needs above those of all other beings. As Taylor (1981) explains:

> We may have responsibilities with regard to the natural ecosystems and biotic communities of our planet, but these responsibilities are in every case based on the contingent fact that our treatment of those ecosystems and communities of life can further the realization of human values and/or human rights. We have no obligation to promote or protect the good of nonhuman living things, independently of this contingent fact.
>
> (p. 198)

Generally speaking, two approaches seeking to move beyond a human-centered view have emerged—biocentrism and ecocentrism.[8] In the space below, I review both of these paradigms with an eye towards understanding how they figure into the question of rights for nonhuman entities.

Taylor (1981) writes that a biocentric ethic features four central tenets: (1) Earth's community of life includes humans, who are members of this group on the same basis as nonhuman entities; (2) ecosystems consist of a complex array of interconnected parts whose proper biological functioning relies on a system of mutual interdependence; (3) every organism within the ecosystem serves a purpose of its own and pursues its own good, which involves striving to achieve its full biological potential; and (4) humans are not superior to any other biological organism (pp. 206–7). Taylor refers to this approach as a "life-centered theory of environmental ethics" (1981, p. 197). The logic of this theory proceeds as follows:

P_1: Humans have moral obligations toward things that possess inherent worth;
P_2: All living things possess inherent worth;
P_3: All members of the biotic community are living;
C: Humans have moral obligations toward members of the biotic community.

Two concepts relevant to biocentrism require further explanation—intrinsic value and inherent worth. Intrinsic value is defined in relational terms. An entity possesses intrinsic value "insofar as some person cherishes it, holds it dear or precious, loves, admires, or appreciates it for what it is in itself, and so places intrinsic value on its existence" (Taylor, 1984, p. 151).[9] Such entities may be living (i.e., plants) or non-living (i.e., places). Key here is the idea that humans bestow intrinsic value on an entity by virtue of their interest in it. Inherent worth, on the other hand, refers to the "value something has simply in virtue of the fact that it has a good of its own" (Taylor, 1984, p. 151). Only living beings have inherent worth. Yet despite the cognizable inherent worth of nonhuman living beings, only humans have rights. Moral rights can only be held by moral agents, and humans are the only moral agents. However, humans have duties towards entities that possess inherent worth. To simplify, humans have a duty to protect nonhuman members of the biotic community like plants and animals not because of the rights they possess, but rather because of their inherent worth.

Interestingly, Taylor briefly contemplates the question as to whether a machine might satisfy tenet three above, specifically the extent to which an entity might possess a good of its own that it strives to pursue. In particular, he limits the discussion to only those machines that are goal-directed and self-regulating. However, Taylor notes that the functions and goals of such machines are programmed by humans, so they do not possess their own sense of good. As such, artificial entities like intelligent machines might have intrinsic value, but not inherent worth. He ultimately arrives at the kind of functionalist argument described in the context of artificial intelligence (AI) in the introduction to this book: "When machines are developed that function in the way our brains do, we may well come to deem then proper subjects of moral consideration" (Taylor, 1981, p. 200, n. 3). This qualification represents something of an odd departure from Taylor's main argument, as it places greater emphasis on the cognitive basis for a being's sense of good than on the position of an entity within an ecosystem. Perhaps it is just "biological chauvinism" (Manzotti & Jeschke, 2016, p. 180) masquerading as "biotic egalitarianism" (Taylor, 1984, p. 156, n. 9).

Writing 15 years after the publication of his famous law review article, Stone (1987) addresses the extent to which machines might be said to have a good of their own. Anticipating the kinds of issues now facing AI ethicists, he observes that "as programs grow more complex, the outputs will become less foreseeably a product of any programmer's original intentions" (Stone, 1987, p. 29). This speculative intervention in the logic from which inherent worth is derived suggests that intelligent machines might be owed respect, if not rights, by moral agents such as humans.

Spitler (1982) critiques biocentrism on the grounds that despite Taylor's best efforts, it is impossible to escape an anthropocentric perspective that prioritizes "human values and experiences" (p. 256). Instead of pretending to consider nature on equal footing with humans while denying the practical implications of such a view,[10] we should "view other forms of life as precious without necessarily declaring them equally precious to human life" (Spitler, 1982, p. 260). In a rejoinder to Spitler's article, Taylor (1983) clarifies that while the biocentric approach may originate from human beliefs, it does not necessarily follow that human interests should take priority above those of other species. In fact, Taylor contends that the charge of inescapable anthropocentrism is not significant, as it conflates the substance, practical implementation, and psychological acceptance of biocentrism. I return to this tension later in the section on critical environmental law.

Nearly two decades later, Wetlesen (1999) attempts to revive and revise biocentrism. Rejecting the argument from marginal cases proffered by animal rights theorists and Taylor's premise that all living things possess inherent worth, the author assigns gradual moral status for nonpersons on the basis of their inherent properties and the degree to which they share similarities with humans. The properties that establish a rubric for determining inherent value include being subject-of-a-life, consciousness, and sentience.[11] The greater the presence of these properties in a given entity, the stronger the case for inherent value, the higher the moral status ascribed to the entity, and the more duties we have towards it. In short, Wetlesen eschews a relational approach to moral status in favor of a properties-based approach that only considers the ethical treatment of organic beings.

In some ways, Wetlesen's work represents a step backwards for advocates of a biocentric ethic. He anchors an entity's inherent value to the degree to which it satisfies a delimited set of properties that closely approximate human characteristics. He denies equal moral status to all living things, electing instead to erect a hierarchy that positions humans above nonhumans. Finally, he promotes biological individualism over ecological holism, essentially seeing the trees, but not the forest. This last move is somewhat surprising given Taylor's occasional nods to a more ecologically oriented biocentric perspective. For instance, Taylor (1981) writes that "the good of the population" is dependent upon maintaining "a coherent system of genetically and ecologically related organisms" (p. 199). In Taylor's biocentric outlook, all living beings are interdependent, and the good of individual biotic communities cannot be achieved in the absence of a stable ecosystem. These tropes—community, ecology, population, system—offer a foundation for an ecocentric environmental ethic, which I turn to next.

Ecocentrism stands as a "natural progression from biocentrism" (Torrance, 2013, p. 405). Whereas a biocentric ethic focuses on individual biological organisms within an ecosystem and observes that although humans are the parties capable of identifying inherent worth in other entities, they are not more morally significant than other living beings, an ecocentric ethic treats the whole ecosystem as the unit of ethical concern and values the preservation of ecological integrity and stability.[12] This "nature-centered" (Hoffman & Sandelands, 2005, p. 144) approach is concisely articulated in Leopold's (1966) famous maxim: "A thing is right when it tends to preserve the integrity, stability, and beauty of the biotic community. It is wrong when it tends otherwise" (p. 262).[13]

Among philosophical efforts to develop an ecocentric ethic, perhaps none is better known than the intellectual movement of deep ecology. As an explicitly normative environmental philosophy, deep ecology promotes "[r]ejection of the man-in-environment image in favour of the *relational, total-field image*" (Naess, 1973, p. 95; emphasis in original). As opposed to shallow ecology, an anthropocentric view that merely seeks to reduce pollution and consumption in the service of improving the lives of people in highly industrialized countries, deep ecology urges a transition from biotic to *biospherical* egalitarianism. The biosphere (also called *ecosphere*) consists of "individuals, species, populations, habitat, as well as human and non-human cultures" (Naess, 1995d, p. 68). In the biosphere, humans are viewed as part of the natural environment, not separate from it. Instead of viewing the world as an assortment of islands, deep ecology perceives life as a web in which humans are but a single strand (Capra, 1995, p. 20).

Although deep ecology has multiple roots, its normative system finds particular compatibility with non-Western spiritual traditions such as Buddhism (Naess, 1995d, p. 79). More specifically, the idea that humans are not the center of the ethical world but rather part of a larger assemblage of ongoing processes is reflected in the Buddhist idea of non-substantiality of the self (*anattaa*) or simply "non-self" (Sponberg, 2000). This dynamic ontology dislocates and destabilizes the self, making it possible and perhaps necessary to identify with *all* beings. Indeed, the Japanese philosopher Dōgen, whose work has served as a source of inspiration for deep ecologists, interprets the scope of Buddha-nature as encompassing both sentient *and* non-sentient beings (Curtin, 2014, p. 269). For some Buddhists, the notion that the world consists of independent entities is merely an illusion. Only one who comes to the realization that everything is united in a web of relations can be considered an enlightened being (*bodhisattva*) (James, 2000, p. 361). This realization results in the *"enlargement of one's sphere of identification,"* which compels us to recognize that all life shares the same fate "not because it *affects* us but because it *is* us" (Fox, 1984, p. 200; emphasis in original).[14]

Importantly, deep ecologists interpret "life" more broadly than do advocates of the biocentric view. In this ecocentric platform, "life" is construed "in a more comprehensive non-technical way" to include organic but "non-living" entities such as ecosystems, landscapes, and rivers (Naess, 1995d, p. 68). All forms of life possess intrinsic value[15] that exists irrespective of their practical usefulness for or appreciation by humans (Naess, 1995d, p. 69). Fox (1990) suggests that life as it is conceived in deep ecological circles concerns a "symbiotic human attitude" directed "not only toward all *members* of the ecosphere but even toward all identifiable *entities* or *forms* in the ecosphere" (p. 116; emphasis in original). To this point Fox adds the important qualification that adherence to deep ecology does not necessarily require adoption of an ecological variant of hylozoism, the view that "[e]very physical object is alive" (Rucker, 2008, p. 364). This brief observation will resurface in the discussion of New Materialism that appears later in this chapter.

As suggested above, deep ecology operates under a gestalt ontology in which "the whole is greater than the sum of its parts" (Naess, 1995a, p. 241). This ontology rejects the idea that humans are separate from the environment in favor of a holistic view that sees everything as interrelated (Wu, 2019, p. 439). Relationships, not individual qualities, provide the foundation for ethical consideration. Crucially for deep ecologists, "there is no firm ontological divide in the field of existence" (W. Fox, 1984, p. 196). However, this ontological orientation is not without its tensions.

For one, deep ecology shifts between holism and individualism. James (2000) questions the coherence of an ethic that affords pride of place to both holism and the intrinsic value of individuals. Deep ecology appears to feature this flaw given its simultaneous recognition of the biosphere, forms of life within it, and relationships among members of it. But James reconciles this apparent contradiction by arguing that deep ecology advances a kind of two-step "thing-centered" holism (2000, p. 367) in which deep experiences of nature include both perceived interconnectedness and acknowledgment of the integrity of individual beings.

Another tension involves whether or not rights have a place in deep ecology. While some anthropocentric environmental ethicists reject rights for nonhumans out of hand (i.e., Norton, 1982),[16] Naess (1995a) holds that all living beings, from the mole to the mountain, possess a right to live and flourish. When the interests of living beings come into conflict, however, two main factors help determine which entity's interests should be awarded priority over the other—nearness (i.e., how close in terms of culture, space, species, and time) and vitalness (i.e., how crucial to a being's existence). To these factors Naess adds a third (which is really a subtype of the first)—felt nearness (i.e., how close in terms of emotional and physical distance). This last factor "determines our capacity to strongly identify with a certain kind of living being, and to suffer when they suffer" (Naess, 1995b, p. 224).

Writing about the concept of the ecological self in Buddhism, Sponberg (2000) pushes back against the ecocentric project and its extension of rights to the natural world. The problem, as the author sees it, is that this effort applies a "Western notion of a permanently fixed sense of selfhood" to determining moral consideration and the possession of rights (Sponberg, 2000). By contrast, Buddhism seeks to reveal the moral consideration of other entities through a transformation in how one sees herself along the path to enlightenment. For example, by engaging in the Buddhist meditative practice of progressively "generating the emotion of loving kindness" towards one's self and eventually all beings, one can develop beyond environmentally destructive tendencies without resorting to an affirmative declaration of rights for living entities (Sponberg, 2000). This practice (*mettaa*) is reflected in the deep ecological idea of "self-realization," which offers that "[t]hrough identification, [people] may come to see that their own interests are served by conservation, through genuine self-love, the love of a widened and deepened self" (Naess, 1995c, p. 229).

Finally, deep ecology fails to completely evade the charge of anthropocentrism. Aside from the fact that this ethical movement is designed and implemented by humans, the manner in which nonhuman entities gain recognition as identifiable forms (i.e., self-realization) or conflicts of interests among living beings are resolved (i.e., felt nearness) relies on and privileges human experience. Even in its earliest manifestations, deep ecology explicitly noted the exalted status of humans and their interests relative to the interests of nonhumans. For instance, Naess (1973) writes that the quality of life humans enjoy "depends in part upon the deep pleasure and satisfaction we receive from close partnership with other forms of life" (p. 96). For Fox (1984), the value inherent to relations with other entities derives from an organism's "capacity for richness of experience" (p. 199), itself a marker for assessing the degree to which nonhuman forms possess sufficiently human-like cognitive abilities. These tensions indicate the persistence of unresolved (or even unresolvable) aspects of an environmental ethic that seeks to capture a larger set of entities than that included in biocentrism's ethical universe.

Another ecocentric approach, this time drawing on deep ecology and transpersonal psychology, deserves mention here. In an effort to capitalize on the strengths of deep ecology and address some of its weaknesses (mainly that the fundamental tenets of the ethical worldview can be used to justify both anthropocentric and ecocentric positions), Fox (1990) proposes a *transpersonal ecology*. Seizing upon deep ecology's notion of self-realization, transpersonal ecology infuses ecocentric ethics with psychological insights regarding the self. Importantly for our purposes, this approach entails three bases for one's identification with other entities—cosmological, personal, and ontological. Cosmological identification, with its origins in non-Western worldviews, emerges from the "deep-seated realization of the fact that we and all other entities are aspects of a single unfolding reality" (W. Fox, 1990, p. 252). This realization can occur through the empathic adoption of any number of different cosmologies, each of which may be capable of demonstrating the unity that exists among all beings in the world. Personal identification results from physical or emotional contact with concrete or abstract entities that are meaningful to us. Such entities are psychologically bound to our own identity; we feel hurt when they suffer, we feel happy when they are well. Finally, ontological identification, a concept whose ineffable character relates closely to the Buddhist pursuit of enlightened consciousness, stems from the "realization of the fact that things are" (Fox, 1990, p. 250; emphasis omitted). More precisely, ontologically based identification suggests that all things that exist are set in brilliant contrast against a void of nothingness. As such, the environment renders itself

> not as a mere backdrop against which our privileged egos and those entities with which they are most concerned play themselves out, but rather as just as much an expression of the manifesting of Being (i.e., of existence per se) as we ourselves are.
>
> (W. Fox, 1990, p. 251)

These modes of identification, distinguished by their respective emphases on reality, identity, and being, offer different ways of relating the self to the world that lies beyond us. Importantly, they specify how we might conceive an ethical outlook that acknowledges the role that humans play in determining the boundaries of the moral circle without privileging human desires and wellbeing in the process.

To briefly summarize the environmental ethics explored here, it helps to ask several questions designed to elucidate some of the meaningful differences between them. First, *what is the basis for ethical consideration?* For biocentrists, it is being an organic living thing, as all living things possess inherent worth. For ecocentrists, it is relationships among members of the biotic community, which includes organic but non-living entities and other forms. Second, *what is the ontological orientation?* For biocentrists, it is individualism; entities are valued separately although they are acknowledged to pursue their own good within the same community of life. For ecocentrists, it is holism;[17] the biosphere is irreducible to its constituent parts, and the idea that the world comprises separate entities is effectively dissolved. Third, *what is the status of humans relative to the environment?* For biocentrists, humans are morally equivalent to other living beings and, despite their individual value, all are bound together in the fate of the ecosystem. For ecocentrists, humans are not distinct from any other members of the ecosphere; they are merely "knots in the biospherical net or field of intrinsic relations" (Naess, 1973, p. 95). Finally, *what place is there, if any, for rights?* For biocentrists, under a life-centered environmental ethic, rights involve "acknowledging a diversity of competing interests of living entities, human and non-human" (Emmenegger & Tschentscher, 1994, p. 579).[18] However, while nonhuman entities such as plants and animals do not possess moral rights, they nonetheless may qualify for legal rights, which would merely concretize the protection they are entitled to by virtue of their inherent worth (Taylor, 1984, p. 218). For ecocentrists, all living and non-living beings in the ecosphere possess the right to live and "pursue their own evolutionary destinies" (W. Fox, 1984, p. 194), although an ecological reading of Buddhism would emphasize transforming how one views one's self and relations with others through the pursuit of enlightenment instead of adopting the Western notion of individual rights (Sponberg, 2000).

Considering the above, what insights might biocentric and ecocentric environmental ethics bring to bear on the discussion regarding the extension of rights to nonhuman inorganic entities?[19] First, both approaches hold differing views regarding the kinds of entities afforded ethical consideration and the extent to which rights might be extended to them. While biocentrism privileges organic beings that pursue their own good (i.e., humans, plants, and animals), it leaves room for inorganic intelligent machines so long as they operate in a way that approximates the functioning of the human brain. However, such machines, like plants and animals, would not be entitled to moral rights. Rather, they would be subject to ethical treatment that does not place human concerns above theirs (provided they were sufficiently intelligent). Further, although both living and non-living beings may be said to hold intrinsic value, possession of such value does not translate into moral rights, which are extended only to humans.

Ecocentrists, on the other hand, offer some rhetorical cover for including inorganic beings in the ethical universe, as suggested by recognition of "identifiable entities or forms in the ecosphere" (W. Fox, 1990, p. 116) and the Latourian observation that there is "no firm ontological divide in the field of existence" (W. Fox, 1984, p. 196). Unlike biocentrism's individualist, life-centered approach to ethics, ecocentrism's ethical system maintains a focus on the "relational, total-field image" (Naess, 1973, p. 95; emphasis omitted) that attempts to acknowledge the whole of the ecosystem along with its constituent parts. Ecocentrism is also more amenable to the idea of rights for nonhuman inorganic entities, although not without some creative interpretation. Naess quite clearly indicates that all living things possess the right to live. Yet, his approach to resolving conflicts among competing interests pertaining to the exercise of this right opens the door for considering other kinds of entities. In particular, the concepts of nearness, vitalness, and felt nearness can be applied to matters involving virtually any beings due to their emphasis on the physical needs of entities and the psychological connection between certain entities and humans. For instance, in accordance with these concepts, an entity that (1) holds cultural significance, (2) could cease to function in the face of a threat to its operation, and (3) is sufficiently life-like to inspire humans to become emotionally invested in its continued existence might qualify for the right to live by *de facto*. As described above, Fox's transpersonal ecology offers three different pathways through which humans might experience the kind of identification with another entity advanced by felt nearness—cosmological, personal, and ontological. To summarize, although at first blush ecocentrism appears to limit ethical consideration to only organic living beings, its holistic ontology and experiential epistemology create a space capable of recognizing a wider range of entities, including inorganic technological forms. Such entities might, at a minimum, satisfy the conditions necessary to deserve the right to live based on the extent to which humans identify with them.

Second, biocentrism and ecocentrism characterize entities in different ways that lead to alternative forms of personhood, resulting in their eligibility for different kinds of rights. For biocentrists, all living beings (i.e., individuals and communities) possess inherent worth because they have a good of their own (i.e., realizing their/its biological potential) that can be impacted positively or negatively by moral agents (i.e., humans). Possession of such a good qualifies an entity for the status of moral patient (Lee, 1999, p. 141). Sentience and consciousness, traits normally associated with psychological personhood (Vincent, 1989, p. 703), are not required for moral patiency (Taylor, 1981, p. 200). Further, nonhuman living entities do not possess rationality, responsibility, or the capacity to articulate their interests—characteristics necessary for demonstrating moral personhood (Vincent, 1989, p. 701). Their actions cannot be judged on moral grounds, and they cannot be said to deliberately violate the rights of others (Taylor, 1984, p. 157). However, moral agents (i.e., humans) have moral obligations towards nonhuman moral patients (i.e., plants and animals) that find concrete expression in legal rights. Therefore, for all intents and purposes, while humans are presently the only entities that might qualify for moral personhood, other living beings may appropriately be categorized as moral patients that can enjoy legal, though not moral, rights. Yet, nonhuman inorganic entities might fall under the purview of moral personhood to the extent that they can demonstrate a capacity for acting rationally and determining their own interests. Theoretically speaking, then, intelligent machines could conceivably be considered moral persons under a biocentric approach, assuming their cognitive abilities exceed those of plants and animals. Still, this would not entitle them to moral rights, since only moral agents possess such rights, and only humans can be moral agents.

For ecocentrists, especially advocates of deep or transpersonal ecology, replacing an individualist orientation with a holistic one and focusing on the ways in which entities are identified and humans identify with them highlights relations among, as opposed to the specific properties of, beings. In the biosphere, all living

and non-living beings co-exist, with no entity worthy of more ethical consideration than another. From Naess's concepts of self-realization and felt nearness to
Fox's three modes of identification, ecocentrism clearly emphasizes the importance of relations in an ethical worldview (Coeckelbergh, 2010, p. 216). This
focus on relations helps to overcome the tendency of philosophers to view entities
in isolation when assessing their moral status (Rodogno, 2016, pp. 52–53). The
move from a properties-based evaluation to a relations-based evaluation precludes
attachment to some forms of personhood while invoking another. In particular,
ecocentrism casts aside psychological and moral personhood, while remaining
somewhat agnostic about legal personhood, in favor of relational personhood.
As some have argued, "personhood is always relational" (Fowler, 2018, p. 397),
so establishing that nonhuman entities might enjoy the status of a person on a
relational basis ultimately feeds into the extent to which they can be seen through
the prism of other forms of personhood, with implications for the assignment of
moral or legal rights. Therefore, under an ecocentric orientation, living and non-
living beings alike might satisfy the conditions for moral personhood, but only by
virtue of their relations with other entities within the biosphere.

Third, while biocentrism could arguably adhere to either the will or interest
theory of rights, ecocentrism exclusively appeals to interest theory. To briefly
review, under will theory, only actors with sufficient mental capabilities can
direct others to fulfill or ignore duties owed to them. Under interest theory, rights
are extended to "entities that have interests and whose interests are furthered
by duties" (Kurki, 2017, p. 79). For biocentrists, while machines might operate
in a way that is functionally similar to human cognition, thus affording them a
performatively autonomous capacity to control duties owed to them, individual
organisms might have interests, but these do not translate into moral rights. They
would only be entitled to legal rights at best. Therefore, biocentrism offers modest, hypothetical support for rights of robots under will theory, but despite recognizing the interests of living beings, the approach stops short of extending rights
to them under interest theory. For ecocentrists, the range of entities eligible for
moral consideration in the biosphere includes living and non-living beings that
do not necessarily possess cognitive capacities, and yet "animals and plants have
interests in the sense of ways of realizing inherent potentialities" (Naess, 1995c,
p. 229) that are revealed through our interactions with them. As such, ecocentrism
effectively rejects any argument for rights under will theory while offering space
for the rights of both living and non-living entities under interest theory via the
three modes of identification and felt nearness.

To summarize, whereas biocentrism extends legal rights to nonhuman organic
and inorganic beings but moral rights only to humans, ecocentrism contends that
all organic, living and non-living identifiable entities in the biosphere deserve the
right to live. However, conflicts over the protection of this right are negotiated on
the basis of identification, suggesting that living and non-living beings in certain
contexts may qualify for relational personhood, which serves as a launchpad for
the designation of moral and legal rights. Clearly then, biocentrism and ecocentrism both support at least legal rights for nature, while only ecocentrism offers a

potential avenue for inorganic non-living entities such as intelligent machines to possess moral *or* legal rights. In the next section, I begin the transition from the philosophical to the practical by examining the ontological scope of the environment and the kinds of entities subject to legal consideration within it through the lens of critical environmental law.

Critical environmental law in the Anthropocene: (Re)defining nature and legal persons

The deficiencies of environmental law that frustrate its ability to address environmental challenges have long been known to observers. In the context of the United States, environmental law is still relatively new compared to other areas of the law; it cuts across many areas of the law without necessarily establishing itself as a distinctive area of practice; and its implementation is subject to political and jurisprudential whims that ignore the need for timely adjudication of environmental disputes (Tarlock, 2004, p. 217). Despite advances in the science of ecology (Brooks, 1991, p. 2) and robust discussions among environmental ethicists (Hirokawa, 2002, p. 226), environmental law has remained relatively unmoved. Attention to these defects has inspired a number of reformatory programs flying under various banners—wild law (Cullinan, 2003), Earth jurisprudence (Koons, 2008), critical environmental law (Philippopoulos-Mihalopoulos, 2011), green legal theory (M'Gonigle & Takeda, 2013), legal ecology (Paloniitty, 2015), and Earth system law (Kotzé & Kim, 2019), among others. Two of the main endeavors crucial to the implementation of these programs involve (1) defining the boundaries of nature and its position relative to humans; and (2) determining what constitutes a legal person. In this section, I focus mainly on the responses offered by critical environmental law, which is "an environmental law that exerts a radical critique of traditional legal and ecological foundations, while proposing in their stead a new, mobile, material and acentric environmental legal approach" (Philippopoulos-Mihalopoulos, 2013, p. 863).

The task of redesigning environmental law to become more responsive to the conditions presently facing humanity and the natural world has enjoyed renewed vigor in light of scholarly attention to the Anthropocene.[20] The arrival of this geological epoch presents a moment for reflecting on the ways in which modern systems of law and governance have failed to prevent the current environmental crisis. In particular, the Anthropocene calls upon us to question whether an anthropocentric worldview is sustainable, given the havoc it has wrought on the Earth and its inhabitants (Vermeylen, 2017, p. 138). This era of heightened ecological awareness also poses a kind of paradox. On the one hand, humans acknowledge the unique impact they have had on the environment and that any effort to meaningfully revise the status quo will require the demotion of human interests. On the other hand, this situation reifies the centrality of humans among members of the living order as the only beings capable of coming to this realization. More concisely, humans are ethically indistinguishable from other entities while also exceptional for their ability to articulate this perspective. This paradox constitutes the Anthropocene dilemma. While critical environmental law interrogates these tensions and disrupts conventions inherent to Western systems of law, the Anthropocene creates the intellectual space necessary "for an opening up of hitherto prohibitive epistemic 'closures' in the law, of legal discourse more generally, and of the world order that the law operatively seeks to maintain" (Kotzé & Kim, 2019, p. 3).

As arguably the most radical of the solutions proposed above, critical environmental law presents intellectually demanding approaches to carrying out the two endeavors while remaining sensitive to the particular context of the Anthropocene. First, *how should we characterize the relationship between humans and the environment?* In a critical reading of environmental law, the Cartesian notion of a definitive split between humans and nature is unsustainable. In its place emerges "hybrid connections between the human, the natural, the spatial, the artificial, the technological" (Philippopoulos-Mihalopoulos, 2011, p. 19). These connections form assemblages that are gathered in alternate combinations depending on the situation. As such, environmental law becomes less fixated on applying a static definition of the environment and more like an autopoietic (i.e., self-creating) system such as an amoeba, which expands and contracts to include/exclude parts of the environment based on the circumstances. Perhaps uncomfortably for some, under such a system "[t]he environment … remains uncharted and unknowable" (Philippopoulos-Mihalopoulos, 2011, pp. 26–27).

Instead of construing humans as actors and nature as the unwitting spectator (i.e., climate change) or vice versa (i.e., natural disasters), all potential entities are marked by vulnerability, a situation in which one is thrown into the world, exposed to it, and aware of it. This awareness does not require consciousness but rather some sense of potential harm that might befall an entity. All such entities find themselves not in the center but in the middle of an immanent space, a place where hierarchy among beings has been replaced by a surface containing all beings at once. Everything that appears on the surface (i.e., all that is) is revealed, rendering it vulnerable. Indeed, "[a]s soon as one is present, one is vulnerable" (Philippopoulos-Mihalopoulos, 2013, p. 859). Thus, for critical environmental law, the ethical priority previously assigned to certain types of entities by virtue of their inherent worth, intrinsic value, or properties is eliminated completely in favor of the acknowledgment that all assemblages in this open ecology are vulnerable.

The Anthropocene highlights the artificiality of human/nonhuman and nature/culture divides, suggesting instead "an ontology of continuous connection between bodies" (Philippopoulos-Mihalopoulos, 2017, p. 132). The resulting continuum is acentric and multi-agentic, a "manifold, full of fissures and planes" (Philippopoulos-Mihalopoulos, 2017, p. 123). As Philippopoulos-Mihalopoulos (2017) argues, "[t]he main find of the Anthropocene is that our presence on the Earth necessarily includes our 'environment', whether 'natural' or otherwise. We are always in an assemblage with the planet. A body is an assemblage of various conditions and materialities" (pp. 125–6). The notion that humans are in some sense separate from the Earth they inhabit is facile and glosses over the various ways in which the ontological boundaries have been effectively blurred in this geological era.

Several philosophers advance arguments that engage with the radical conception of the environment described above. One area of philosophical inquiry involves the extent to which abiotic objects (i.e., non-natural artefacts) might be considered part of the environment. While some maintain that "[a]rtefacts are ... discontinuous from nature" (Bryson, 2018, p. 17), others contend that there are degrees of artifacticity that depend on the kind of materials used and the amount of human effort expended in the design and creation of an object (Lee, 1999, pp. 52–53). The former perspective may be characterized as ontological dualism (i.e., viewing some entities as inferior to others), whereas the latter is an example of ontological dyadism (i.e., recognizing and celebrating differences among entities). Yet, a more radical take that resonates with critical environmental law suggests that the ontological division between the natural and the artefactual established by the importance assigned to human intention ignores the unanticipated effects generated by artefacts. Therefore, we should replace "nature" with "reality" and accept that artefacts are continuous with natural entities (Vogel, 2015, p. 105). This perspective shares with critical environmental law a rejection of Cartesian dualism and a preference for assemblages.

A related and broader philosophical discussion focuses on the boundaries of nature itself. The etymology of the word *nature* suggests that the term refers to a space filled with life that is nevertheless separate from humans (Merleau-Ponty, 2003, p. 3). As indicated earlier in this section, the environment that is the (tortured) object discussed by environmental ethicists is almost without exception a natural one. Rarely do non-living entities, let alone inorganic ones, find a home in these environs. But the present phase of modernity, a period many refer to as the Anthropocene, calls into question the strict delineation between the natural and the non-natural. In this crucial time, "[i]t no longer makes sense to consider nature as the backdrop against which human activities evolve" (Hey, 2018, p. 351). If, for instance, humanity has had such a profound impact on nature that it is no longer possible to distinguish the natural environment from the built one, the former may now be said to include the latter. As Vogel (2015) argues, *contra* McKibben (1989), "[t]he distinction between humans and nature ... depends on a philosophically and biologically untenable dualism that forgets that human beings themselves are part of nature and instead treats them as exceptional creatures who somehow transcend the natural" (p. 24). The extent of human influence on the environment is now so undeniable that we have moved from an existence predicated on "being-in-the-world" (to use Martin Heidegger's phrasing) to one of "being-in-the-technological-world" (to borrow from Hans Jonas) (Tavani, 2018, p. 12). The biosphere (Naess, 1995d) has been effectively subsumed by a continuously rupturing contingent reality. The point here is that the environment, far from being some idealized location free from human intervention, is not (and perhaps in the annals of human history never really was) a purely natural one, but rather a built one, one *we* built. In line with critical environmental law, it does not make sense to think of humans as separate from a *natural* environment.

Indeed, "the environment we encounter and live within is always already a *built* environment" (Vogel, 2015, p. 58; emphasis added). Such a conception of the environment includes human *and* nonhuman, natural *and* artefactual entities. The practical application of this interpretation can be seen in the definition of environmental justice used by the National Association of County and City Health Officials, which explains how "'environment' includes the ecological, physical, social, political, aesthetic, and economic environment" (NACCHO, 2019).

Second, *who or what counts as a legal person?* In mainstream Western conceptions of law, a "legal person" often refers to a nonhuman entity that possesses legal personality, such as a corporation,[21] while the term "natural person" usually denotes a human being (Grear, 2013, p. 78). The law treats the latter as the quintessential rational subject, the fulcrum around which objects like the environment must rotate. Throughout history, the privileged position occupied by the natural person has permitted legal discrimination *among* members of the human species (i.e., intra-species hierarchy) and *against* animals, ecosystems, and other nonhuman entities (i.e., inter-species hierarchy) (Grear, 2015, p. 230). The elevated status conferred upon humans by contemporary legal systems is predicated on the idea that we alone possess dignity and reason, and thus law must work to advance the human good (Pietrzykowski, 2017, p. 49).

A critical environmental law responsive to insights provided by the Anthropocene challenges these foundational assumptions on three accounts. First, it forces legal systems to reflexively confront their systematic biases. Critical environmental law recognizes the fragility of concepts like "natural person," which is exposed as a mere construct favoring "a white, property owning, acquisitive, broadly Eurocentric masculinity" (Grear, 2015, p. 236). Acknowledgment of such restrictive and power-laden criteria for designating legal relevance marks the first step towards opening up legal systems to new possibilities. Second, it erodes the notion of a clear nature/culture divide and offers in its place a void where new perspectives can take hold. Freed from the shackles of Western "juridical humanism" (Pietrzykowski, 2017, p. 49), law can entertain alternative views like those found in Indigenous cosmologies. For example, one Indigenous view asserted by Amazonian peoples considers both animals and humans to be people. In this perspective, "the form of species is just merely a clothing or an 'envelope' hiding an internal human form" (Vermeylen, 2017, p. 146).[22] In brief, critical environmental law's radical openness presents an opportunity for conceptualizing legal personhood in diverse ways. Finally, it de-centers and de-individualizes the legal person through a fresh understanding of materiality urged by the Anthropocene. In this geological epoch, nature has been hybridized, making it difficult (if not impossible) to understand where humans end and nature begins (Arias-Maldonado, 2019, p. 51). The Anthropocene demonstrates that the fortunes of both people and the planet rise and fall together; any semblance of autonomy has been lost (Vermeylen, 2017, p. 153). The complexification of hitherto static concepts like "man" and "environment" is assisted by New Materialism, which posits that "if everything is material inasmuch as it is composed of physiochemical processes, nothing is reducible to such processes, at least as conventionally understood. For materiality is always something more than 'mere' matter" (Coole & Frost, 2010, p. 9). One implication of this philosophical movement is that, *contra* Descartes' rigid conditions of materiality, there no longer exists a meaningful distinction between organic and inorganic matter.[23] Both are imbued with vitality and agency by virtue of their material constitution. As all kinds of *lively* matter are intrinsically and irreversibly intertwined with other forces and systems, the capacity for agency necessarily extends beyond the human form (Coole & Frost, 2010, p. 9). Another implication is that environmental legal subjects might now refer "not to individual bodies, subjects, experiences or sensations, but to assemblages of human and non-human, animate and inanimate, material and abstract, and the affective flows within these assemblages" (N. J. Fox & Allred, 2015, p. 406). The result is that the traditional subject of environmental law (i.e., the human agent) finds itself "de-centered" and "repositioned" to a place within, though not atop, ecologies consisting of "interspecies dependencies" (Grear, 2017, p. 93).

The above arguments culled from critical environmental law, the Anthropocene, and New Materialism suggest that the contemporary Western conception of a legal person is not only outmoded, but also environmentally hazardous. As Grear (2015) observes, "[l]aw's dominant construction of legal personhood—and law in general, including international environmental law—are thus unresponsive— at a fundamental level—to the ethical implications of the vulnerable embodied bio-materiality of the living order" (p. 241). A radically ecological envisioning of legal personhood would situate environmental subjects in the middle, not the center, of an open ecology where all entities endure a similar vulnerability, while simultaneously acknowledging that humans bear a special responsibility for their unique role in causing planetary destruction (Arias-Maldonado, 2019, p. 56). With their newfound openness to alternate perspectives (i.e., Indigenous and otherwise) inspired by critical environmental law's call for reflexiveness and ecological epistemology, modern legal systems could accommodate both living and non-living, organic and inorganic, natural and artefactual legal persons as situations dictate.

To conclude, critical environmental law and cognate intellectual departures from mainstream legal and philosophical thought offer two key innovations relevant to the topic of rights for nonhuman entities. First, nature includes both the natural and built environments. In the age of the Anthropocene, it does not make sense to speak of a division between the two. Humans act upon and are continuous with the environment, which includes natural and non-natural beings. Second, legal personhood can extend to nonhuman, non-living, inorganic, and/or artefactual entities. The agentic capacity of all matter coupled with the preference for recognizing persons as inevitably bound up in assemblages opens up a complex and evolving basis for determining what counts as a legal person. Alternative perspectives on personhood, such as those inscribed in Indigenous cosmologies, are welcomed. For example, the Lakota saying at the beginning of this chapter, *mitákuye oyás'į* ("we are all related"), advances a *relational* form of personhood that extends to both humans and nonhumans (Posthumus, 2017, p. 385). However, no perspective is viewed as inherently superior to another. In the following section, I shift over to the world of law-in-action in order to examine cases in which nature has been found to possess rights and compare the extent to which the reasoning employed by judges comports with arguments made in environmental ethics and critical environmental law.

The rights of nature in court

As mentioned towards the beginning of this chapter, while the ideas underlying the RoN are not new, their successful adjudication in courts around the world is a fairly recent development. To be sure, cases involving the RoN are not just about bringing nature into the fold of human legal systems in order to protect it; they are ideological battlegrounds for debating the merits of ontologies that prescribe different kinds of relations among humans and nonhumans (Youatt, 2017, p. 41). As such, these cases hold the potential to expose anthropocentric systems of law that have facilitated environmental destruction through rapacious economic development (Calzadilla & Kotzé, 2018, p. 399). In this section, I examine how the RoN have been interpreted and justified in a few celebrated cases in order to understand the extent to which jurisprudence in this area resonates with arguments from environmental ethics and critical environmental law on the rights of nonhuman entities. The objective is to determine how, if at all, case law on the RoN might inform the discourse on rights for technological entities. To this end, I analyze the legal reasoning employed in four RoN cases involving rivers[24] and glaciers in order to distill the evidence and rationale(s) used to arrive at their respective rights-affirming decisions.

In *Wheeler c. Director de la Procuraduria General Del Estado de Loja* (Wheeler),[25] plaintiffs sought to enjoin continuation of a project to widen the Vilcabamba-Quinara road, which was causing rocks and other materials to be deposited in the nearby Rio Vilcabamba in southern Ecuador. Failing to adhere to best practices of environmental management, the developers "had not carried out an environmental impact assessment, secured planning permits for the construction, or planned for the disposal of debris that would inevitably occur" (Daly, 2012, p. 63). As a result of the excavation and displacement of debris, the Rio Vilcabamba was effectively narrowed, reducing the flow of the river while increasing its speed (Greene, n.d.). This alteration of the landscape caused erosion and subsequent flooding, negatively impacting people living along the river. Complainants Richard Frederick Wheeler and Eleanor Geer Huddle argued that the development project violated the RoN, but more specifically the rights of the Rio Vilcabamba itself. On March 30, 2011, the provincial Court of Justice in Loja found in favor of Wheeler and Huddle, granting a constitutional injunction on the basis that the RoN had been disregarded and finding the provincial government liable for damages.

The legal reasoning applied by the court in the world's first-ever vindication of the RoN relied on constitutional law, impacts on future generations, and the relationship between humans and nature. The court specifically cited Article 71 of the Ecuadorian Constitution, which describes the rights of Nature or "Pacha Mama"[26] (Ecuador Const., tit. II, ch. 7, art. 71). Immediately following the constitutional reference, the decision went on to explain how "injuries to Nature are 'generational injuries' which are such that, in their magnitude have repercussions not only in the present generation but whose effects will also impact future generations" (Daly, 2012, p. 64). Finally, in support of the previous point, the court quoted at length a speech by former president of the National Constituent Assembly Alberto Acosta, who himself cites Aldo Leopold's ecocentric maxim mentioned earlier in this chapter. Also in his speech, Acosta makes the following arguments in support of advancing "Earth democracy":[27]

a) individual and collective human rights must be in harmony with the rights of other natural communities on earth;
b) ecosystems have the right to exist and follow their own vital processes;
c) the diversity of life expressed in nature is a value in itself;
d) ecosystems have their own values that are independent of their usefulness for the human being;
e) the establishment of a legal system in which ecosystems and natural communities have an inalienable right to exist and prosper situational to nature at the highest level of value and importance.[28]

Importantly, Acosta also contended that "[t]he human being is a part of nature, and [we] must prohibit human beings from bringing about the extinction of other species or destroying the functioning of natural ecosystems" (Daly, 2012, p. 64). That the aforementioned speech was included directly in the text of the decision (i.e., not dicta) speaks to the authoritative role that Acosta's ideas played in the court's reasoning.

The court's interpretation of the RoN, driven mainly by Acosta's speech, offers a few important insights regarding the extension of rights to natural non-human entities in the Ecuadorian context. First, injuries to nature are wrong not only because they violate or frustrate Pacha Mama, but also because they affect humans in the present and future. Second, while human rights should not conflict with the RoN, the latter should be given priority. Third, rights are bestowed upon whole ecosystems or natural communities, not individual natural entities. Fourth, humans are part of nature, although they possess the ability to act upon it as well.

These points reflect some of the ideas discussed in the literatures on environmental ethics and critical environmental law. In terms of ethics, the Ecuadorian RoN appear to express more of an ecocentric outlook than a biocentric one. Ecosystems possess a value of their own outside of that which might be assigned by humans, and natural communities are considered in their totality, not on a strictly individual basis. In terms of critical environmental law, the court recognized that humans should not retain their position of privilege as the central subject of environmental law, but at the same time, some of the reasoning suggested that human impacts remain crucial to analyses of alleged violations of RoN. Although there seems to be little room in the decision for moving beyond natural or organic nonhuman entities, the de-centering of humans and the de-individualization of legal persons observed in the decision promote a moderately critical stance towards environmental law. Finally, while the court does observe that humans are entwined with nature, the ruling does not attempt to redefine the boundaries of the environment.

In *Center for Social Justice Studies et al. v. Presidency of the Republic et al.* (Atrato River),[29] Indigenous and Afro-descendent communities, under the auspices of an *acción de tutela*,[30] sought to halt illegal mining and logging activities along the Atrato River in the biodiverse Chocó district of Colombia. The claimants argued that these activities, specifically the dumping of harmful chemicals into the river, were causing pollution that threatened water quality, aquatic life, agricultural production, and the lives of children from ethnic communities. These impacts, according to the plaintiffs, animated concerns about protecting "the fundamental rights to life, health, water, food security, a healthy environment, the culture and the territory of the active ethnic communities."[31]

After the claim was initially denied by several government entities due to standing issues, a failure to pursue all available avenues for redressing the grievances, and a rejection of the allegation that illegal activities were being conducted along and in the river, the Delegate Ombudsman for Constitutional and Legal Matters took up the case on behalf of the affected communities. Following unsuccessful efforts to proceed with the *acción de tutela* at the Administrative Tribunal of Cundinamarca and State Council, the case was reviewed by the Sixth Chamber of Revision, which added to the scope of government agencies implicated in the complaint. The Constitutional Court of Colombia considered the merits of the revised claim.

After concluding that the plaintiffs did indeed have an admissible *acción de tutela*, the court turned to discuss environmental issues. Despite recognizing the principle of human dignity as a "superior value" within Colombia's legal order,[32] the court examined the constitutional relevance of environmental protection. Referencing articles pertaining to collective rights and state obligations, the court emphasized "constitutional guarantees for the general welfare and productive and economic activities of the human being to be carried out in harmony and not with sacrifice or to the detriment of nature."[33] Importantly, the decision articulated what the justices saw as a synergistic relationship between environmental (i.e., the "Ecological Constitution") and cultural matters (i.e., the "Cultural Constitution").[34] The former focus urges safeguarding the environment for the sake of people and nature in their own right, while the latter highlights the inextricable link between nature and culture, leading to the conservation of biodiversity for the ways in which it protects the vitality of traditional cultures.

The court alternately reviewed three theoretical approaches to identifying the importance of nature in the Colombian constitutional system—anthropocentrism, biocentrism, and ecocentrism. Citing its own case precedent, the court found specific support for an ecocentric approach in which "nature is not conceived only as the environment and surroundings of human beings, but also as a subject with its own rights, which, as such, must be protected and guaranteed."[35] This conclusion was bolstered by the court's discussion of biocultural rights, which it defined as the rights that ethnic communities have to administer and exercise autonomous guardianship over their territories—according to their own laws and customs—and the natural resources that make up their habitat, where their culture, their traditions and their way of life are developed based on the special relationship they have with the environment and biodiversity.[36]

Biocultural rights resonate with an ecocentric approach to constitutional interpretation because they observe a "profound unity between nature and the human species" that respects the role that Indigenous relationships with nonhuman natural entities play in fostering biodiversity.[37] In support of clarifying its obligations to protect biocultural rights, the court enumerated several international legal instruments that have been incorporated into Colombian law—ILO Convention 169 on Indigenous and Tribal Peoples, the Convention on Biological Diversity, the UN Declaration on the Rights of Indigenous Peoples, the American Declaration on the Rights of Indigenous Peoples, and the UNESCO Convention for the Safeguarding of Intangible Cultural Heritage.[38] It was on the basis of developments regarding biocultural rights at the international level that the court found nature to be "a subject of rights."[39] By explicitly adopting an ecocentric perspective, the court endeavored to deliver justice to nature while acknowledging the culturally significant relationship that the environment has with humans.

The *Atrato River* case provides three crucial insights related to the rights of nonhuman entities in Colombia. First, legal complaints may be considered justiciable when human interests, especially the rights of ethnic communities, are affected, not necessarily because nature itself suffers violations of its own rights. Second, the RoN are supported by homegrown case precedent, environmental and cultural aspects of the Colombian Constitution, international treaties to which Colombia is party, and biocultural rights. Third, the court has demonstrated a willingness to endorse an ecocentric approach to constitutional interpretation in which environmental and biocultural rights are fundamentally conjoined and mutually reinforcing.

The above summary suggests that Colombia's treatment of the RoN invokes aspects of environmental ethics and critical environmental law pertinent to the ontological centrality of humans and the kinds of subjects that might qualify for legal personhood. Clearly the court's active embrace of ecocentrism demonstrates a jurisprudential preference for more holistically ecological thinking, although the origins and final decision of the case suggest that anthropocentrism may not be so easily divorced from Colombian legal proceedings. Nature may have rights, but their violation might not inspire legal action until humans become affected as well. As far as parallels with critical environmental law, through its full-throated advocacy for biocultural rights, the court opened the door for de-centering humans and de-individualizing legal subjects in the Colombian system. Although it did not directly entertain the possibility of non-natural legal subjects, the court's emphasis on acknowledging Indigenous ontologies left space for the introduction of entities that hold certain cultural significance, irrespective of their material composition. This line of argumentation could conceivably support a critical expansion of the environment that includes the built domain, but the judgment itself does not make this point explicitly.

In the State of Uttarakhand, the High Court decided two cases in the same year that advanced the RoN in India. In *Mohd Salim v. State of Uttarakhand & Others* (Ganges and Yamuna),[40] the plaintiff alleged that the governments of Uttarakhand and Uttar Pradesh had failed to cooperate with the Central Government when they did not appoint representatives to the Ganges Management Board, an institution set up to protect the river and its tributaries (Safi, 2017). The establishment of such an inter-state agency was deemed necessary given that the river ecosystems had been severely degraded due to pollution and unlawful activities (O'Donnell, 2018, p. 136).

In rendering its decision, the High Court tied together three main arguments. First, the Ganges and Yamuna rivers are sacred for Hindus, so the threat to their existence "requires extraordinary measures."[41] Second, these rivers contribute to the health and well-being of "communities from mountains to sea."[42] Third, the concept of legal person (or what the judges call a "juristic person") can evolve according to the needs of society.[43] Thus, in order to protect religiously significant entities that serve important spiritual and physical functions, they need to be brought into the fold of the legal system as legal persons.

In support of the latter argument, the justices cited Indian case law demonstrating that nonhuman entities of religious importance can be considered legal persons when doing so serves "the needs and faith of society."[44] Interestingly, the High Court also added two qualifications regarding the concept of legal personhood in the Indian context. First, the decision asserts that "juristic" persons are similar to natural persons in the sense that they hold rights and obligations.[45] The main difference between the two is that the former act through an intermediary, which is enabled through innovations in Indian law like the concept of "representative standing" (Cunningham, 1987, p. 499). Second, the judges held that since developments in society naturally lead some entities to transform from a mere "fictional personality" into a "juristic person," legal personhood could be conferred upon "any entity," even "objects or things."[46] Given the important role that they play in Indian society, therefore, the Ganges and Yamuna rivers are considered "legal persons/living persons."[47]

In *Lalit Miglani v. State of Uttarakhand & Others* (Glaciers),[48] High Court advocate Miglani "explicitly sought to extend legal personhood to all other natural objects in the State of Uttarakhand, including the Gangotri and Yamunotri glaciers that provide headwaters for the Ganges and Yamuna rivers" (O'Donnell, 2018, p. 136). But before discussing the ruling in the *Glaciers* case, it is necessary to provide some additional background information. The petition filed in this case stemmed from an earlier complaint made by Miglani alleging "gross failure" by governmental authorities to prevent pollution of the Ganges river (Shivshankar, 2017). Finding in favor of the plaintiff, the High Court issued an order recognizing the right to clean water, the polluted state of the Ganges river, and its significance to the Hindu population.[49] This order also included, *inter alia*, a mandate to establish an inter-state Council among all Indian states whose jurisdiction reaches parts of the Ganges.[50] The High Court's ruling did not, however, resolve the issue raised by the petitioner.

Following the disposition of the 2016 order, the High Court decided the *Ganges and Yamuna* case, which granted legal personhood to rivers and treated juristic persons as equivalent to natural/living persons. In light of this ruling and under the auspices of a miscellaneous application, Miglani sought to have the High Court declare "the Himalayas, Glaciers, Streams, Water Bodies etc. as legal entities as juristic persons at par with pious rivers Ganga and Yamuna."[51] In order to render its judgment on this request, the same panel of two judges who decided the *Ganges and Yamuna* case, Judges Rajiv Sharma and Alok Singh, explored U.S. and Indian Supreme Court jurisprudence,[52] environmental literature, environmental science, international environmental law, legal scholarship, and, importantly, New Zealand's (2014) *Te Urewera Act*, which gave the Urewera National Park the status of a legal entity on the basis of its spiritual significance for the Tuhoe, an Indigenous group (Rodgers, 2017, p. 272).

Drawing from these varied sources of inspiration, the High Court arrived at several conclusions regarding the legal status of the Gangotri and Yamunotri glaciers and their accompanying rights. First, bodies of water possess an "intrinsic right not to be polluted."[53] Second, polluting natural entities like air, forests, glaciers, and rivers is "legally equivalent" to harming a natural person.[54] Third, natural entities "have a right to exist, persist, maintain, sustain and regenerate their own vital ecology system."[55] Fourth, humans and their natural surroundings constitute a "unified and … indivisible whole."[56] The High Court then reiterated the purpose and definition of a juristic person first articulated in the *Ganges and Yamuna* case, holding that "the Himalayan Mountain Ranges, Glaciers, rivers, streams, rivulets, lakes, jungles, air, forests, meadows, dales, wetlands, grasslands and springs are required to be declared as the legal entity/legal person/juristic person/juridicial person/moral person/artificial person for their survival, safety, sustenance and resurgence."[57]

Unlike the ruling in the *Ganges and Yamuna* case, the decision in the *Glaciers* case made direct mention of climate change and Mother Earth in ways that evoke the concerns expressed in writings on law in the Anthropocene. The judges acknowledged that climate change poses an existential threat to natural entities on land and sea,[58] and argued in favor of using legal rights to protect Mother Nature given our moral duties owed to future generations.[59] These arguments can be seen as working synergistically in the context of the Anthropocene, marrying scientific assessments of planetary conditions with Indigenous worldviews that promote the RoN (Knauß, 2018, p. 703).

Viewed in tandem, these two cases demonstrate the evolution of legal thinking about the rights of nonhuman entities occurring in Indian jurisprudence. First, while natural entities may hold certain religious importance that justifies their need for protection, they also possess intrinsic rights related to their own ecological wellbeing. Second, humans and nature comprise a unified whole, and their fortunes are tied together. Third, legal personhood is a concept that is determined in accordance with societal needs, so any entity could potentially be declared a juristic person. Fourth, legal persons are equivalent to natural persons in terms of the rights they possess and the need to prosecute violations of those rights, although the former may be represented in legal proceedings by a designated human.

Together, the *Ganges and Yamuna* and *Glaciers* cases relate to ideas found in environmental ethics and critical environmental law regarding the human–environment relationship and the extension of legal personhood to nonhuman entities. Humans are part of the environment, and natural beings such as bodies of water provide physical and spiritual sustenance to the Indian people. This perspective reflects an ecocentric intention, but in practice it is moderated by the emphasis placed on preserving nature on the basis of protecting religious values, the performance of ecosystem services that benefit people, and the nod towards providing an environment of equal quality for future generations. However, given the understanding that humans might be the only entities capable of recognizing their unique responsibility to act as stewards of the environment by utilizing institutions like the law, the hints of anthropocentrism may not be so problematic after all. On the issue of legal personhood, the Indian cases offer perhaps the most progressive vision discussed in this section. In both rulings, the judges held that legal personhood has been and should be determined by the needs of society. While prior Indian jurisprudence has awarded legal status to religious idols, today the extent of environmental degradation that has befallen rivers and glaciers warrants their consideration as legal persons as well. According to this theory, literally any entity could be granted the status of legal person in light of societal justifications. Therefore, the approach to legal personhood exhibited in the Indian cases comes closer than those described before them to manifesting a view in line with critical environmental law, although the extent to which it de-centers humans and de-individualizes legal subjects is not strongly pronounced. In addition, while no direct effort was made to incorporate the built environment into nature, the fact that non-natural entities could become legal persons might make this ontological move unnecessary as a practical matter.[60]

A brief comparison of the cases reviewed in this section illustrates how the RoN have been adjudicated across different jurisdictions (see Table 4.1).

The cases analyzed above suggest a few points regarding judicial interpretation of the RoN and the future of RoN jurisprudence. First, complaints tend to be reactive (as opposed to proactive) and involve alleged injuries to both the environment and people. At this stage in their infancy, the RoN appear to be activated when humans recognize and decide to challenge failures to protect the environment. Second, the sources of legal reasoning implicitly draw upon global norms related to RoN and explicitly on a combination of national, foreign, and international law, and cultural ideas specific to each given jurisdiction. As such, the findings summarized here concur with those of Kauffman and Martin (2018), who show that RoN norm construction involves multiple forms of influence at varying levels of governance. Third, rulings regarding the RoN have generally appealed to an ecocentric environmental ethic, sometimes directly. Fourth, each case spoke the language of critical environmental law, albeit with different levels of fluency. While

Table 4.1 Comparison of rights of nature cases

Case	Origin of Complaint	Source(s) of Legal Reasoning	Ethical Approach	Evidence of Critical Environmental Law	Legal Subjects Considered
Wheeler	Specific development project harming nature and humans	Domestic law and Indigenous ideas	Ecocentric	Yes; de-centers humans and de-individualizes legal subjects, but does not redefine nature	Nonhuman *natural* entities
Atrato River	Specific development project harming nature and humans	Domestic law and Indigenous ideas	Ecocentric	Yes; de-centers humans and de-individualizes legal subjects; but does not redefine nature	Nonhuman *natural* entities
Ganges and Yamuna	General environmental degradation harming nature and human values	Domestic law and religious beliefs	Ecocentric	Yes; expands definition of legal person, but does not redefine nature	Nonhuman entities of *any kind*
Glaciers	General environmental degradation harming nature and human values	Domestic, foreign, and international law, and religious beliefs	Ecocentric	Yes; expands definition of legal person, but does not redefine nature	Nonhuman entities of *any kind*

some decisions actively de-centered humans in the realm of contemporary legal systems and de-individualized legal subjects in ways that reflect an ecocentric ethical orientation, others expanded the definition of a legal person to include entities beyond humans and nature. Although none of the cases redefined the boundaries of nature to include the built environment, expanding the kinds of entities worthy of legal personhood on the basis of societal need welcomes nonhuman non-natural entities into the vulnerable space in the middle of the open ecology.

Conclusion

Taken together, what insights about the rights of artefactual entities can be gleaned from the discussions about environmental ethics, critical environmental law, and RoN cases? First, ecocentrism, considerably more so than biocentrism, can support a progressive extension of rights beyond the biosphere. Deep ecology in particular recognizes both human and nonhuman cultures, and is compatible with non-Western spiritual traditions like Buddhism, which have a dynamic ontology similar to that which is expressed in critical environmental law. Further, deep ecology recognizes identifiable entities or forms in the biosphere, which could include those that are non-living in the natural sense, but nonetheless important to traditional cultures. An emerging area of scholarship applies Indigenous ideas to modern technology. Analysts writing on the subject argue that Indigenous epistemologies can serve as the basis for acknowledging "an extended 'circle of relationships' that includes the non-human kin—from network daemons to robot dogs to artificial intelligences (AI) weak and, eventually, strong—that increasingly populate our computational biosphere" (Lewis et al., 2018, p. 2). This Indigenously derived notion of "making kin with the machines" (Lewis et al., 2018) fits well with transpersonal ecology's concept of cosmological identification. In short, ecocentrism and its cognate ethical programs of deep ecology and transpersonal ecology are flexible enough to accommodate non-Western worldviews that utilize alternative epistemologies to identify entities natural and non-natural that are part of the environment, warrant ethical consideration, and might deserve rights.

Second, critical environmental law and its attention to the crises of the Anthropocene advocate for a radical revision to environmental law that disrupts conventional notions of nature and legal persons. In the Anthropocene, the Cartesian divide between humans and nature is exposed as intellectually bankrupt, an ontological heresy. Instead, it is not only more accurate but also prudent to view the natural and built environments as enveloped by the broader *physical* environment (Gellers, 2016). This space is an open ecology in which entities of all kinds are present and thus vulnerable. However, the open ecology is unstable and autopoietic, drawing in different combinations of entities and relationships as situations dictate. Here, legal persons are no longer defined by the extent of their moral agency. New Materialism has shown agency to exist in all matter and act as a process, so the properties deemed necessary for an entity to possess agency cannot serve as the criteria by which moral or legal personhoods are established. Instead, it is preferable to refer to assemblages, not agents. Further, critical environmental law is highly sensitive to the imperialistic and dominating tendencies of contemporary Western legal systems, so in an effort to decolonize environmental law non-Western worldviews must be considered as valid as those that subjugated them, although no approach to legal thought should be privileged above another. Therefore, the range of entities that lay claim to the status of "legal person" and its attendant rights is highly dependent upon the context in which conflicts arise and the type of perspective adopted. In practical terms, a legal approach that recognizes Indigenous kinship with forms of technology like social robots would have to take seriously the possibility that such intelligent machines would constitute artefacts within the environment that may qualify for relational (and later psychological/moral and legal) personhood and moral or legal rights.

Third, cases involving the RoN demonstrate how evolving global norms are adapted in different national contexts, clarifying how legal systems (re)interpret nature and legal personhood in light of traditional and diffused ideas. Interestingly, judges in each of the cases reviewed above rendered their respective decision using an ecocentric, as opposed to biocentric, ethical orientation. However, the rulings also reflect the kind of deeply embedded anthropocentrism that reveals itself in the form of the Anthropocene dilemma. That is to say, they simultaneously call for the elevation of nature's interests to a level at least on par with that of humans while overlooking the facts that in each case, the complaint arose from alleged harm to humans, and that humans are presently the only entities capable of bringing these issues before courts. In addition, all four cases exhibit some degree of consonance with the objectives of critical environmental law, albeit in slightly different ways. While the number of cases examined here are insufficient to arrive at generalizable conclusions about the usefulness of RoN jurisprudence in expanding the list of potential legal subjects, a more tentative assessment would be that while the Latin American cases offer grounds for granting legal personhood to nonhuman *natural* entities, the Indian cases unequivocally support recognizing *artefactual* entities, such as intelligent machines, as legal persons, provided society has a need for doing so.

Demonstrating how extant moral and legal paradigms can be adjusted to accommodate technological beings will require two additional moves—proposing a new framework for determining personhoods and prescribing an ethic with room for recognizing robots. These remaining tasks consume the pages of the final chapter, which we turn to next.

Notes

1 See Ecuador Const., tit II, ch. 7.
2 See Law 071 of the Rights of Mother Earth (2010) and Framework Law 300 of Mother Earth and Integral Development for Living Well (2012).
3 The Community Environmental Legal Defense Fund lists Stone's 1972 article as the first of many "key moments in the development of the movement for the Rights of Nature" (CELDF, 2019). At the time of writing, the article has over 2,000 citations according to Google Scholar.

4 *Sierra Club v. Morton*, 405 U.S. 727 (1972).
5 Ibid., at 741–742.
6 Ibid., at 743.
7 For an in-depth discussion of the role played by CELDF in the writing and passage of the Tamaqua Borough ordinance, see Kauffman and Martin (2018, pp. 53–54).
8 Admittedly, this three-pronged characterization of environmental ethics (anthropocentrism, biocentrism, ecocentrism) greatly oversimplifies the totality of philosophical positions present in the field. However, given the focus of this book and the intent to communicate crucial elements of environmental ethics to those new to the area, I chose to restrict the diversity of views to a more easily digestible and still not inaccurate classification system. For those interested in understanding the range of philosophical orientations flying under the banner of environmental ethics, see Fox (1990, pp. 22–35).
9 This derivation of intrinsic value differs from Taylor's earlier argument, which held that if an entity was a living being, "the realization of its good is something *intrinsically valuable*" (Taylor, 1981, p. 201; emphasis in original).
10 Specifically, Taylor's antagonist proposes a hypothetical situation in which shooting one's neighbor would be viewed as morally equivalent to "swatting a fly or stepping on a wild flower" (Spitler, 1982, p. 260). Taylor addresses this objection at length in his rejoinder, arguing that under a biocentric outlook, it will always be wrong to kill wild animals and plants, and doing so would be "as much a wrong as killing or harming a human," but this does not mean that there are no valid reasons for killing a fly or wildflower (Taylor, 1983, p. 242). On the contrary, any such action must be justified in terms of moral reasoning that exceeds the nature of the wrongful act.
11 Wetlesen notably uses the term "inherent value" instead of Taylor's "inherent worth," and the preference is not a trivial one. The author notes how "intrinsic value" and "inherent value" are often used interchangeably, but he opts for the latter because it suggests that an entity's moral status is determined by "properties ... internal to the nature of the subject" (Wetlesen, 1999, p. 290). For Taylor, an entity's moral status is a function of its possession of a good of its own, not its internal properties. This is an important distinction.
12 Some have argued that the emphasis on achieving ecological stability or equilibrium is misguided because it creates the impression that ecosystems existing in certain states are more valuable than others. Instead, wildness is preferred as an element of an ecocentric ethic because of its ambivalence towards the type of ecological restoration pursued. See Hettinger and Throop (1999).
13 Fox (1990, p. 177) maintains that Leopold's land ethic and Naess's ecocentric ethic are distinguished only in terms of scope, referring to the former as ecosystemic (i.e., limited to the local ecosystem) and the latter as ecospheric (i.e., pertaining to the whole ecosphere).
14 But see James (2014), who counters that "[t]he good Buddhist treats nature well ... not because she believes she is 'one' with the natural world, but because she has, through practice, come to develop certain virtues of character. She treats nature well, that is, because she is compassionate, gentle, humble, mindful, and so on, not just in relation to her fellow humans, but in her dealings with all things" (p. 112).
15 Note that Naess's phrase "intrinsic value" is synonymous with Taylor's notion of "inherent worth" mentioned earlier.
16 More specifically, Norton (1982) claims that there exists a disjuncture between the interests of individual organisms and the integrity of the ecosystem. Granting rights to members of an ever-increasing moral circle is merely a way of attempting to capture all victims of harm caused by environmental degradation. But a comprehensive environmental ethic can accomplish the same heavy lifting without resorting to rights. Therefore, extending rights to the whole of the environment is unnecessary at best and paralyzing at worst. He concludes that "it is exactly this individualistic charac-

ter of rights which makes the attempt to generate an environmental ethic from rights unpromising—I would even say impossible" (Norton, 1982, p. 33).

17 But see Emmenegger and Tschentscher (1994, pp. 577–578), who argue that holism is not synonymous with ecocentrism; and McShane (2014), who contends that individualism and holism are variants of biocentrism.

18 Biocentrists appear to have focused less on the question of rights than their ecocentric counterparts. Legal scholars have notably expended more effort in this endeavor that environmental philosophers. Curiously, one legal analyst suggests that the RoN represent a biocentric (not ecocentric) outlook. See Borràs (2016).

19 Interestingly, antipathy towards the introduction of rights into environmental ethics have come from both biocentric and ecocentric quarters. Rolston III (1993), a biocentrist, makes a Hohfeldian critique of the RoN, observing the absence of legitimate claims and entitlements among beings in the wild. As such, the very idea of rights applied to nonhumans and natural entities is "comical" and "inappropriate" (Rolston III, 1993, p. 256). Duncan (1991), an ecocentrist, similarly takes issue with the RoN, but for different reasons. First, perhaps ironically, extending rights to the environment may have the effect of reifying the Cartesian separation between humans and nature. Second, Western legal systems, whose rights are based on a philosophical commitment to individualism, cannot intellectually accommodate holistic approaches to legal protection. This is all to say that neither biocentrists nor ecocentrists support rights for nature with universal approbation.

20 The irony of naming such a period of contemplation the *Anthro*pocene is not lost on Grear (2017), who asserts that the Western, human-centered focus of the term "simply extends the logics of Eurocentric human exceptionalism and methodological individualism—the self-same logics that gave rise to the Anthropocene crisis itself" (p. 79).

21 Some terminological differences can be observed across the literature on legal personhood. For instance, Kurki (2017) refers to legal persons as "artificial persons" (p. 74).

22 Even this non-Western cosmological form of anthropomorphism maintains a kind of anthropocentrism. Critical environmental law, by contrast, promotes a posthuman ontology that seeks to de-center humans entirely (even those dressed as animals). I thank Anna Grear for bringing this to my attention.

23 For a more in-depth discussion of matter in the context of New Materialism, see Coole (2010).

24 It is not immediately clear why so many RoN cases and laws focus primarily on rivers, although Cano-Pecharroman (2018) suggests two potential explanations: "The first ruling recognizing the rights of nature was regarding a river, and the existence of this previous jurisprudence could have provided foundations for other judges to rule in the same way. The nature of rivers as a distinct mass of water elapsing across terrain with a quasi-permanent shape and presence may make it easier to legally define a river as an 'object' that can become a 'subject' with rights" (p. 6).

25 *Wheeler c. Director de la Procuraduria General Del Estado de Loja*, Juicio No. 11121-2011-0010 (2011) ('Wheeler'), available at https://elaw.org/system/files/ec.wheeler .loja_.pdf.

26 Although the phrase "Pacha Mama" is often inaccurately translated to mean "Mother Nature" (Blaser, 2014, p. 51), it is more properly interpreted as a "philosophy of life" that involves "living in harmony with nature, co-existing with it, caring for it, and allowing for its regeneration to provide for the upcoming generations" (Cano-Pecharroman, 2018, p. 6).

27 Translation author's own with the assistance of Google Translate.

28 *Wheeler*, at 3–4.

29 *Center for Social Justice Studies et al. v. Presidency of the Republic et al.*, Judgment T-622/16, Constitutional Court of Colombia (2016) ('Atrato River'), available at http:/ /www.corteconstitucional.gov.co/relatoria/2016/t-622-16.htm.

30 An *acción de tutela* (or "guardianship action") is a legal mechanism provided under Article 86 of the Colombian Constitution that streamlines the process of bringing a claim involving the violation of fundamental rights by public authorities (*¿Qué es la Acción de Tutela?*, 2017).
31 *Atrato River Case*, translated by the Dignity Rights Project, available at https://delawar elaw.widener.edu/files/resources/riveratratodecisionenglishdrpdellaw.pdf.
32 Ibid., at 26.
33 Ibid., at 30.
34 Ibid., at 31.
35 Ibid., at 34.
36 Ibid., at 35.
37 Ibid., at 37.
38 Ibid., at 39.
39 Ibid., at 98.
40 *Mohd Salim v State of Uttarakhand & Others*, WPPIL 126/2014, High Court of Uttarakhand (2017) ('Ganges and Yamuna'), available at https://elaw.org/system/files/ attachments/publicresource/in_Salim__riverpersonhood_2017.pdf.
41 Ibid., at 4.
42 Ibid., at 11.
43 Ibid., at 10.
44 Ibid., at 7.
45 Ibid., at 10.
46 Ibid.
47 Ibid., at 11.
48 *Lalit Miglani v State of Uttarakhand & Others*, WPPIL 140/2015, High Court of Uttarakhand (2017) ('Glaciers'), available at https://indiankanoon.org/doc /92201770/.
49 *Lalit Miglani v State of Uttarakhand & Others*, WPPIL 140/2015, High Court of Uttarakhand (2016), available at https://indiankanoon.org/doc/189912804/.
50 Ibid., at 73.
51 *Glaciers*, at 2.
52 The citations to U.S. Supreme Court jurisprudence focused mainly on the principle of *parens patriae*, which involves situations where the state has a quasi-sovereign interest in representing individuals who cannot represent themselves for one reason or another. See *Snapp & Son, Inc. v. Puerto Rico ex rel. Barez*, 458 U.S. 592 (1982).
53 Ibid., at 61.
54 Ibid.
55 Ibid.
56 Ibid.
57 Ibid., at 63.
58 Ibid., at 61.
59 Ibid., at 61, 62.
60 As a coda to the *Ganges and Yamuna* case, the Uttarakhand government challenged the High Court's ruling and the Indian Supreme Court overturned the decision, agreeing with the State that the declaration on the rights of the rivers was "legally unsustainable" (BBC, 2017).

References

Akchurin, M. (2015). Constructing the Rights of Nature: Constitutional Reform, Mobilization, and Environmental Protection in Ecuador. *Law and Social Inquiry*, *40*(4), 937–968.

Arias-Maldonado, M. (2019). The "Anthropocene" in Philosophy: The Neo-material Turn and the Question of Nature. In F. Biermann & E. Lövbrand (Eds.), *Anthropocene Encounters: New Directions in Green Political Thinking* (pp. 50–66). Cambridge University Press.

BBC (2017, July 7). India's Ganges and Yamuna Rivers Are "Not Living Entities." *BBC News*. Retrieved from https://www.bbc.com/news/world-asia-india-40537701.

Biggs, S. (2017, November 1). *Ponca Nation of Oklahoma to Recognize the Rights of Nature to Ban Fracking*. Movement Rights. Retrieved from https://www.movement rights.org/ponca-nation-of-oklahoma-to-recognize-the-rights-of-nature-to-ban-fra cking/.

Blaser, M. (2014). Ontology and Indigeneity: On the Political Ontology of Heterogeneous Assemblages. *Cultural Geographies, 21*(1), 49–58.

Borràs, S. (2016). New Transitions from Human Rights to the Environment to the Rights of Nature. *Transnational Environmental Law, 5*(1), 113–143.

Borrows, J. (1997). Living between Water and Rocks: First Nations, Environmental Planning and Democracy. *University of Toronto Law Journal, 47*(4), 417–468.

Boyd, D. R. (2017). *The Rights of Nature: A Legal Revolution That Could Save the World*. ECW.

Brooks, R. O. (1991). A New Agenda for Modern Environmental Law. *Journal of Environmental Law and Litigation, 6*, 1–38.

Brown, K. (2018, November 14). Ecuador's Indigenous March Over 600km to Demand an End to Mining. *Al Jazeera*. Retrieved from https://www.aljazeera.com/news/2018/11/e cuador-indigenous-march-600km-demand-mining-181115011033618.html.

Bryson, J. J. (2018). Patiency Is Not a Virtue: The Design of Intelligent Systems and Systems of Ethics. *Ethics and Information Technology, 20*(1), 15–26.

Calzadilla, P. V., & Kotzé, L. J. (2018). Living in Harmony with Nature? A Critical Appraisal of the Rights of Mother Earth in Bolivia. *Transnational Environmental Law, 7*(3), 397–424.

Cano-Pecharroman, L. (2018). Rights of Nature: Rivers That Can Stand in Court. *Resources, 7*(1), 1–14.

Capra, F. (1995). Deep Ecology: A New Paradigm. In G. Sessions (Ed.), *Deep Ecology for the 21st Century: Readings on the Philosophy and Practice of the New Environmentalism* (pp. 19–25). Shambhala.

CELDF (2019, June 21). *Advancing Legal Rights of Nature: Timeline*. Community Environmental Legal Defense Fund. Retrieved from https://celdf.org/rights/rights-of-nature/rights-nature-timeline/.

Coeckelbergh, M. (2010). Robot Rights? Towards a Social-Relational Justification of Moral Consideration. *Ethics and Information Technology, 12*(3), 209–221.

Collins, L. M. (2019). Environmental Resistance in the Anthropocene. *Oñati Socio-Legal Series, Online*, 1–19.

Coole, D. (2010). The Inertia of Matter and the Generativity of Flesh. In D. Coole & S. Frost (Eds.), *New Materialisms: Ontology, Agency, and Politics* (pp. 92–115). Duke University Press.

Coole, D., & Frost, S. (2010). Introducing the New Materialisms. In D. Coole & S. Frost (Eds.), *New Materialisms: Ontology, Agency, and Politics* (pp. 1–43). Duke University Press.

Cullinan, C. (2003). *Wild Law: A Manifesto for Earth Justice*. Green Books.

Cunningham, C. D. (1987). Public Interest Litigation in Indian Supreme Court: A Study in Light of American Experience. *Journal of the Indian Law Institute, 29*(4), 494–523.

Curtin, D. (2014). Dōgen, Deep Ecology, and the Ecological Self. In J. B. Callicot & J. McRae (Eds.), *Environmental Philosophy in Asian Traditions of Thought* (pp. 267–290). State University of New York Press.

Daly, E. (2012). The Ecuadorian Exemplar: The First Ever Vindications of Constitutional Rights of Nature. *Review of European Community and International Environmental Law*, *21*(1), 63–66.

Duncan, M. L. (1991). The Rights of Nature: Triumph for Holism or Pyrrhic Victory? *Washburn Law Journal*, *31*(1), 62–70.

Ecuador Const., tit. II, ch. 7, art. 71.

Emmenegger, S., & Tschentscher, A. (1994). Taking Nature's Rights Seriously: The Long Way to Biocentrism in Environmental Law. *Georgetown International Environmental Law Review*, *6*(3), 545–592.

Fowler, C. (2018). Relational Personhood Revisited. *Cambridge Archaeological Journal*, *26*(3), 397–412.

Fox, N. J., & Allred, P. (2015). New Materialist Social Inquiry: Designs, Methods and the Research-Assemblage. *International Journal of Social Research Methodology*, *18*(4), 399–414.

Fox, W. (1984). Deep Ecology: A New Philosophy of Our Time? *The Ecologist 14*(5–6), 194–200.

Fox, W. (1990). *Toward a Transpersonal Ecology: Developing New Foundations for Environmentalism*. Shambhala.

Gellers, J. C. (2016). The Great Indoors: Linking Human Rights and the Built Environment. *Journal of Human Rights and the Environment*, *7*(2), 243–261.

Grear, A. (2013). Law's Entities: Complexity, Plasticity and Justice. *Jurisprudence*, *4*(1), 76–101.

Grear, A. (2015). Deconstructing Anthropos: A Critical Legal Reflection on 'Anthropocentric' Law and Anthropocene 'Humanity.' *Law and Critique*, *26*(3), 225–249.

Grear, A. (2017). "Anthropocene, Capitalocene, Chthulucene": Re-Encountering Environmental Law and Its "Subject" with Haraway and New Materialism. In L. J. Kotzé (Ed.), *Environmental Law and Governance for the Anthropocene* (pp. 77–96). Hart Publishing.

Greene, N. (n.d.). *The First Successful Case of the Rights of Nature Implementation in Ecuador*. Global Alliance for the Rights of Nature. Retrieved October 24, 2019, from https://therightsofnature.org/first-ron-case-ecuador/.

Gudynas, E. (2011). Buen Vivir: Today's Tomorrow. *Development*, *54*(4), 441–447.

Hettinger, N., & Throop, B. (1999). Refocusing Ecocentrism: De-emphasizing Stability and Defending Wildness. *Environmental Ethics*, *21*(1), 3–21.

Hey, E. (2018). The Universal Declaration of Human Rights in "the Anthropocene." *AJIL Unbound*, *112*, 350–354.

Hirokawa, K. (2002). Some Pragmatic Observations About Radical Critique in Environmental Law. *Stanford Environmental Law Journal*, *21*(2), 225–281.

Hoffman, A. J., & Sandelands, L. E. (2005). Getting Right with Nature: Anthropocentrism, Ecocentrism, and Theocentrism. *Organization and Environment*, *18*(2), 141–162.

James, S. P. (2000). "Thing-Centered" Holism in Buddhism, Heidegger, and Deep Ecology. *Environmental Ethics*, *22*(4), 359–375.

James, S. P. (2014). Against Holism: Rethinking Buddhist Environmental Ethics. In J. B. Callicot & J. McRae (Eds.), *Environmental Philosophy in Asian Traditions of Thought* (pp. 99–116). State University of New York Press.

Kauffman, C. M., & Martin, P. L. (2017). Can Rights of Nature Make Development More Sustainable? Why Some Ecuadorian Lawsuits Succeed and Others Fail. *World Development*, *92*, 130–142.

Kauffman, C. M., & Martin, P. L. (2018). Constructing Rights of Nature Norms in the US, Ecuador, and New Zealand. *Global Environmental Politics*, *18*(4), 43–62.

Kingston, L. (2015). The Destruction of Identity: Cultural Genocide and Indigenous Peoples. *Journal of Human Rights*, *14*(1), 63–83.

Knauß, S. (2018). Conceptualizing Human Stewardship in the Anthropocene: The Rights of Nature in Ecuador, New Zealand and India. *Journal of Agricultural and Environmental Ethics*, *31*(6), 703–722.

Koons, J. E. (2008). Earth Jurisprudence: The Moral Value of Nature. *Pace Environmental Law Review*, *25*(2), 263–340.

Kotzé, L. J., & Kim, R. E. (2019). Earth System Law: The Juridical Dimensions of Earth System Governance. *Earth System Governance*, *1*, 1–12.

Kurki, V. A. J. (2017). Why Things Can Hold Rights: Reconceptualizing the Legal Person. In V. A. J. Kurki & T. Pietrzykowski (Eds.), *Legal Personhood: Animals, Artificial Intelligence and the Unborn* (pp. 69–89). Springer.

LaDuke, W. (2019, February 4). *The White Earth Band of Ojibwe Legally Recognized the Rights of Wild Rice. Here's Why*. NationofChange. Retrieved from https://www.nationofchange.org/2019/02/04/the-white-earth-band-of-ojibwe-legally-recognized-the-rights-of-wild-rice-heres-why/.

Lee, K. (1999). *The Natural and the Artefactual: The Implications of Deep Science and Deep Technology for Environmental Philosophy*. Lexington Books.

Leopold, A. (1966). *A Sand County Almanac*. Ballantine Books.

Lewis, J. E., Arista, N., Pechawis, A., & Kite, S. (2018). Making Kin with the Machines. *Journal of Design and Science*, *3*, 5.

Magalhaes, L., & Pearson, S. (2019, August 31). 'I Thought the World Was Ending': What's Fueling the Amazon Rainforest Fires. *Wall Street Journal*. Retrieved from https://www.wsj.com/articles/i-thought-the-world-was-ending-whats-fueling-the-amazon-rainforest-fires-11567224081.

Manzotti, R., & Jeschke, S. (2016). A Causal Foundation for Consciousness in Biological and Artificial Agents. *Cognitive Systems Research*, *40*(C), 172–185.

Margil, M. (2016, September 18). *Press Release: Ho-Chunk Nation General Council Approves Rights of Nature Constitutional Amendment*. CELDF. Retrieved from https://celdf.org/2016/09/press-release-ho-chunk-nation-general-council-approves-rights-nature-constitutional-amendment/.

Maurial, M. (1999). Indigenous Knowledge and Schooling: A Continuum Between Conflict and Dialogue. In L. M. Semali & J. L. Kincheloe (Eds.), *What Is Indigenous Knowledge?: Voices from the Academy* (pp. 59–77). Falmer Press.

McKibben, B. (1989). *The End of Nature*. Random House.

McShane, K. (2014). Individualist Biocentrism vs. Holism Revisited. *Les Ateliers de l'éthique / The Ethics Forum*, *9*(2), 130–148.

Merleau-Ponty, M. (2003). *Nature: Course Notes from the Collège de France* (R. Vallier, Trans.). Northwestern University Press.

Meyer, R. (2016, September 9). The Dakota Access Pipeline, the Standing Rock Sioux Tribe, and the Law. *The Atlantic*. Retrieved from https://www.theatlantic.com/technology/archive/2016/09/dapl-dakota-sitting-rock-sioux/499178/.

M'Gonigle, M., & Takeda, L. (2013). The Liberal Limits of Environmental Law: A Green Legal Critique. *Pace Environmental Law Review*, *30*(3), 1005–1115.

Morgan, R. (2004). Advancing Indigenous Rights at the United Nations: Strategic Framing and Its Impact on the Normative Development of International Law. *Social and Legal Studies*, *13*(4), 481–500.

NACCHO (2019). *Environmental Justice*. National Association of County and City Health Officials. Retrieved from https://www.naccho.org/programs/environmental-health/ju stice.

Naess, A. (1973). The Shallow and the Deep, Long-Range Ecology Movement: A Summary. *Inquiry*, *16*(1–4), 95–100.

Naess, A. (1995a). Ecosophy and Gestalt Ontology. In G. Sessions (Ed.), *Deep Ecology for the 21st Century: Readings on the Philosophy and Practice of the New Environmentalism* (pp. 240–245). Shambhala.

Naess, A. (1995b). Equality, Sameness, and Rights. In G. Sessions (Ed.), *Deep Ecology for the 21st Century: Readings on the Philosophy and Practice of the New Environmentalism* (pp. 222–224). Shambhala.

Naess, A. (1995c). Self-Realization: An Ecological Approach to Being in the World. In G. Sessions (Ed.), *Deep Ecology for the 21st Century: Readings on the Philosophy and Practice of the New Environmentalism* (pp. 225–239). Shambhala.

Naess, A. (1995d). The Deep Ecological Movement: Some Philosophical Aspects. In G. Sessions (Ed.), *Deep Ecology for the 21st Century: Readings on the Philosophy and Practice of the New Environmentalism* (pp. 64–84). Shambhala.

Norton, B. G. (1982). Environmental Ethics and Nonhuman Rights. *Environmental Ethics*, *4*(1), 17–36.

O'Donnell, E. L. (2018). At the Intersection of the Sacred and the Legal: Rights for Nature in Uttarakhand, India. *Journal of Environmental Law*, *30*(1), 135–144.

O'Gorman, R. (2017). Environmental Constitutionalism: A Comparative Study. *Transnational Environmental Law*, *6*(3), 435–462.

Paloniitty, T. (2015). Taking Aims Seriously: How Legal Ecology Affects Judicial Decision-Making. *Journal of Human Rights and the Environment*, *6*(1), 55–74.

Philippopoulos-Mihalopoulos, A. (2011). Towards a Critical Environmental Law. In A. Philippopoulos-Mihalopoulos (Ed.), *Law and Ecology: New Environmental Foundations* (pp. 18–38). Routledge.

Philippopoulos-Mihalopoulos, A. (2013). Actors or Spectators? Vulnerability and Critical Environmental Law. *Oñati Socio-Legal Series*, *3*(5), 854–876.

Philippopoulos-Mihalopoulos, A. (2017). Critical Environmental Law in the Anthropocene. In L. J. Kotzé (Ed.), *Environmental Law and Governance for the Anthropocene* (pp. 117–136). Hart Publishing.

Pietrzykowski, T. (2017). The Idea of Non-Personal Subjects of Law. In V. A. J. Kurki & T. Pietrzykowski (Eds.), *Legal Personhood: Animals, Artificial Intelligence and the Unborn* (pp. 49–67). Springer.

Posthumus, D. C. (2017). All My Relatives: Exploring Nineteenth-Century Lakota Ontology and Belief. *Ethnohistory*, *64*(3), 379–400.

¿Qué es la Acción de Tutela? (2017, December 29). Alviar Gonzalez Tolosa Abogados. Retrieved from https://www.agtabogados.com/blog/que-es-la-accion-de-tutela/.

Reuters (2019, July 28). Thousands Protest Development on Sacred Maori Land in New Zealand. *The Japan Times*. Retrieved from https://www.japantimes.co.jp/news/2019/0 7/28/asia-pacific/thousands-protest-development-sacred-maori-land-new-zealand/.

Rodgers, C. (2017). A New Approach to Protecting Ecosystems: The Te Awa Tupua (Whanganui River Claims Settlement) Act 2017. *Environmental Law Review*, *19*(4), 266–279.

Rodogno, R. (2016). Robots and the Limits of Morality. In M. Nørskov (Ed.), *Social Robots: Boundaries, Potential, Challenges* (pp. 39–56). Ashgate.

Rolston III, H. (1993). Rights and Responsibilities on the Home Planet. *Yale Journal of International Law, 18*(1), 251–279.

Rucker, R. (2008). Everything Is Alive. *Progress of Theoretical Physics Supplement, 173,* 363–370.

Safi, M. (2017, March 21). Ganges and Yamuna Rivers Granted Same Legal Rights as Human Beings. *The Guardian.* Retrieved from https://www.theguardian.com/world/2 017/mar/21/ganges-and-yamuna-rivers-granted-same-legal-rights-as-human-beings.

Shivshankar, G. (2017, April 5). The Personhood of Nature. *Law and Other Things.* Retrieved from https://lawandotherthings.com/2017/04/the-personhood-of-nature/.

Spitler, G. (1982). Justifying a Respect for Nature. *Environmental Ethics, 4*(3), 255–260.

Sponberg, A. (2000). The Buddhist Conception of an Ecological Self. *Western Buddhist Review, 2.* Retrieved from http://www.westernbuddhistreview.com/vol2/ecological_se lf.html.

Stone, C. D. (1972). Should Trees Have Standing? Toward Legal Rights for Natural Objects. *Southern California Law Review, 45,* 450–501.

Stone, C. D. (1987). *Earth and Other Ethics: The Case for Moral Pluralism.* Harper & Row.

Tamaqua Borough. (2006). Ordinance No. 612. Schuylkill County, PA.

Tarlock, A. D. (2004). Is There a There There in Environmental Law? *Journal of Land Use and Environmental Law, 19*(2), 213–254.

Tavani, H. T. (2018). Can Social Robots Qualify for Moral Consideration? Reframing the Question about Robot Rights. *Information, 9*(4), 1–16.

Taylor, P. W. (1981). The Ethics of Respect for Nature. *Environmental Ethics, 3*(3), 197–218.

Taylor, P. W. (1983). In Defense of Biocentrism. *Environmental Ethics, 5*(3), 237–243.

Taylor, P. W. (1984). Are Humans Superior to Animals and Plants? *Environmental Ethics, 6*(2), 149–160.

Te Urewera Act, Pub. L. No. Public Act 2014 No. 51 (2014). Retrieved from http://www .legislation.govt.nz/act/public/2014/0051/latest/whole.html.

Torrance, S. (2013). Artificial Agents and the Expanding Ethical Circle. *AI and Society, 28*(4), 399–414.

Vermeylen, S. (2017). Materiality and the Ontological Turn in the Anthropocene: Establishing a Dialogue Between Law, Anthropology and Eco-Philosophy. In L. J. Kotzé (Ed.), *Environmental Law and Governance for the Anthropocene* (pp. 137–162). Hart Publishing.

Vincent, A. (1989). Can Groups Be Persons? *The Review of Metaphysics, 42*(4), 687–715.

Vogel, S. (2015). *Thinking Like a Mall: Environmental Philosophy After the End of Nature.* MIT Press.

Wetlesen, J. (1999). The Moral Status of Beings Who Are Not Persons: A Casuistic Argument. *Environmental Values, 8*(3), 287–323.

Wu, S. (2019). Ecological Holism: Arne Naess's Gestalt Ontology and Merleau-Ponty's Bodily-Flesh Phenomenology. In A. K. Giri (Ed.), *Practical Spirituality and Human Development: Creative Experiments for Alternative Futures* (pp. 437–453). Springer.

Youatt, R. (2017). Personhood and the Rights of Nature: The New Subjects of Contemporary Earth Politics. *International Political Sociology, 11*(1), 39–54.

5 Rights for robots in a posthuman ecology

> The swarm is, as it were, not oncoming as a distant phenomenon: we are already swarm.
>
> (Matilda Arvidsson, 2020, p. 134)

The revised question posed at the outset of this study was, *under what conditions might some robots be eligible for moral or legal rights?* In this final chapter, I draw together insights from the preceding chapters in order to form a response to this more carefully sharpened line of inquiry. First, I revisit the conceptual map detailed in Chapter Two in order to assess the individual analytical utility of ontological properties and relational mechanisms. Second, I explore how the theory and practice related to animal rights might contribute to the debate over robot rights. Third, I review how environmental ethics, environmental legal scholarship, and case law on the rights of nature (RoN) might inform the discussion on rights for intelligent machines. Fourth, in light of the aforementioned analyses, I present a multi-spectral framework that can be used to determine the conditions under which certain intelligent machines might be eligible for moral or legal rights. Fifth, I propose a praxis-oriented, critically inspired ethic capable of accommodating both organic and inorganic nonhuman entities. Sixth, I work through two hypothetical examples in order to demonstrate how the aforementioned framework and ethic might prove useful in assessing the appropriateness of extending rights to artificial intelligence (AI). Seventh, I suggest areas for further research.

Navigating the conceptual map

The purpose of introducing the conceptual map in Chapter Two was to clarify the relationships between various properties/mechanisms, personhoods, statuses, and incidents/positions. Here I interrogate this heuristic tool further, seeking to extract from it a defensible method for determining the personhood(s) for which an entity might be eligible. In order to accomplish this, I evaluate the individual candidacy of some of the most frequently discussed ontological properties and relational mechanisms. I begin at the top left of the map (i.e., those properties associated with psychological personhood) and finish with the bottom left (i.e., those mechanisms tied to relational personhood).[1] I argue that while relational mechanisms provide a stronger platform than ontological properties for making determinations about moral or legal status, both should be considered in tandem.

Consciousness, as mentioned earlier, is a trait whose relevance to moral concern is commonly debated among AI ethicists. Many argue that only those beings endowed with consciousness are eligible for psychological/moral personhood, moral status, and moral rights. A general interpretation of consciousness suggests that the term refers to "subjective states of sentience and awareness that we have during our waking life" (Searle, 2008, p. 356); that is, consciousness is a capacity for feeling and thought internal to a subject. Another perspective proposes that consciousness can be found not within the brains of agents themselves, but rather in all matter. This idea is known as "panpsychism" (Nagel, 1979; Goff, 2019). An alternate hypothesis suggests that consciousness exists as a causal interaction between agents and their external world (Manzotti & Jeschke, 2016). Still another view adds that there are degrees of consciousness reflecting different levels of cognitive awareness, stages of development, and abilities among species (Turner, 2019, pp. 152–153). As such, it might be more appropriate to talk about consciousness as a phenomenon that varies according to brain activity, age, and type of entity. Absent dispositive scientific evidence, each of these perspectives might be equally plausible. However, our understanding of consciousness is limited by the problem of other minds, making it difficult to verify empirically. Consciousness might also be epiphenomenal to higher-order cognition, making the trait less morally significant than certain behaviors that suggest even nonhuman creatures are capable of mental states (Carruthers, 2005). While considerable evidence points to animals experiencing "at least simple conscious thoughts and feelings" (Griffin & Speck, 2004, p. 5), the question of AI consciousness remains far murkier (Torrance, 2008; Schneider, 2016). For instance, Takayuki Kanda remarked in an interview that "[c]onsciousness requires a full understanding of the world. AI today is still struggling to identify objects." Another interviewee, Yoshikazu Kanamiya, added that "we don't have an appropriate model of consciousness, and without a model it is difficult to talk about whether or not a robot can exhibit consciousness." In short, consciousness is not a strong option for establishing personhood due to unresolved issues regarding how we operationalize it and confirm its presence in other entities.

Intentionality is another quality regarded as important to personhood and moral status. However, there are numerous ways in which intentionality is defined and applied in the literature, making terminological consistency elusive. For instance, scholars have written about folk (Calverley, 2008), functional (Johnson, 2006), and philosophical (Searle, 1980) forms of intentionality, to name but a few of the variants. Even if we subscribe to the philosophical version (perhaps an appropriate choice given the subject matter of this book), which refers to intentionality as "that feature of the mind by which it is directed at or about objects and states of affairs in the world" (Searle, 2008, p. 356), there may be levels of such a capacity that make it necessary but not sufficient for personhood (Dennett, 1976, p. 180). All of the following examples demonstrate intentionality at varying levels of sophistication: a potted plant growing in the direction of sunlight streaming through a window; my dog Shiva leaping on top of her kennel to retrieve her favorite plush toy; a videogame opponent operated by the computer that seeks to defeat its human adversary; and a young woman studying hard for an exam because doing well might help her gain admission to graduate school. Deciphering intentionality requires observers to make inferences about the motivations underlying the actions of other agents. It is therefore at best an indirect, if intuitively appealing, attribute associated with personhood.

Sentience is a property discussed at length in the literature, especially among those writing on animal rights. In general, it is thought to refer to "the capacity to suffer or experience enjoyment or happiness" (Singer, 1974, p. 108). Unlike some other properties, sentience arguably suffers less from definitional ambiguity.[2] However, Gunkel (2012) contends that Singer, the main proponent of sentience as a moral criterion in animal ethics, "conflates suffering and sentience" (p. 84). In fact, Singer's sentience involves "phenomenal consciousness" (i.e., the capacity to feel pain, pleasure, etc.) but not the kind of self-consciousness associated with higher-level internal thought (Torrance, 2008, p. 500). In addition, scholars disagree about the extent to which sentience generates interests (and eventually rights). Singer contends that sentient beings have interests, which compel their moral consideration on equal grounds with humans. By contrast, Fox (1990) draws a comparison between plants and humans to demonstrate that maintaining life is an interest independent from one's capacity for sentience. Therefore, "sentience cannot be considered synonymous with having interests per se. Rather, sentience simply introduces a new class of interests—mentally expressed interests—into the domain of moral considerability" (Fox, 1990, p. 167). On a more practical note, Wise (2013) argues that sentience may prove unhelpful as a benchmark used in animal rights advocacy because "common law judges will accept autonomy, but not sentience, as a sufficient condition for legal personhood" (p. 1286). Complicating matters further, new scientific research on plants "suggests that sentience is a contingent and fluid concept; one that depends upon a constantly changing combination of scientific and cultural assumptions" (Pelizzon & Gagliano, 2015, p. 5). The increasing knowledge about sentience renders its usefulness as a standard for moral or legal consideration more dubious than ever, despite mounting evidence that animals indeed experience suffering (perhaps more so than enjoyment) (i.e., Groff & Ng, 2019). Thinking in terms of robots, sentience requires the capacity for consciousness, which might only come if machines achieve singularity. But Yueh-Hsuan Weng cautioned during an interview that this is "not likely to happen in the near future."

Autonomy is an ontological property associated with personhood and moral agency. Its Greek roots, *autos* (self) and *nomos* (law), indicate that it refers to the capacity to impose law on oneself (Schmidt & Kraemer, 2006, p. 74). A Kantian view suggests that autonomy entails "obedience to the rational dictates of the moral law" (Calverley, 2008, p. 532). In other words, an agent uses reason to decide on a course of action in line with self-imposed moral laws, and thus responsibility for that action may be correctly attributed to it. In the context of AI, autonomy may mean the possession of agency independent from operators, programmers, or users who would otherwise dictate the actions of a technological entity (Sullins, 2006, p. 30). The crucial distinction between these two interpretations is that the former involves qualities possessed by an agent herself and actions traceable to that agent, while the latter emphasizes the degree of causal separation between an artificial agent and its human operator.[3] Calverley (2008) writes that if an action undertaken by an intelligent machine is not traceable to its original human operator, it would be responsible for the ensuing action and therefore deemed sufficiently autonomous to be considered a legal person (p. 533). But the standard for autonomy might be considerably lower among laypersons. As Ryutaro Murayama of robot start-up GROOVE X noted during an interview, people may simply "think" that intelligent machines operate autonomously based on their behavior. Continuing in the legal domain, Wise (2013) finds autonomy to be the main trait underlying dignity, a foundational aspect of human rights. However, the animal rights litigator argues that autonomy, which requires consciousness, is also present in animals, albeit to varying degrees. More directly, consciousness is necessary for autonomy, which is the basis of dignity and therefore rights. The plurality of definitions and interwoven concepts of similarly unsettled meaning make autonomy yet another property that is relevant to both moral and legal inquiries, but difficult to pin down conceptually.

Intelligence is intuitively integral to personhood. Humans have long been designated the intelligent species, while animals were simply (if lovingly) deemed "God's dumb creatures" (Neave, 1909, p. 563). Intelligence clearly figures prominently in the context of technologies alleging to exhibit artificial forms of it. For instance, Turing's (1950) famous imitation game was conceived as a theoretical way of determining whether or not machines could think. In working through his thought experiment, Turing argued that intelligent behavior might be possible through machine learning. Importantly, intelligence was viewed as a process that could be replicated. Searle (1980), on the other hand, proposed his Chinese Room argument as a way of refuting Turing. In this thought experiment, Searle countered that the outputs generated by instructions were not authentic markers of intelligence. For Searle, it is intentionality, not the mere replication of symbols, that demonstrates intelligence.

However, as with many other ontological properties, intelligence lacks a clear definition. Even an "intuitive notion of intelligence may not pick out a single neatly defined cognitive capability" (Shevlin et al., 2019, p. 1). Perhaps tellingly, in the early days of intelligence studies, the construct was determined inductively through the structure and performance of psychometrics (Wagman, 1999, p. 1). Furthermore, intelligence is often understood to mean different things depending on whether one appeals to lay, expert, Western, or non-Western conceptions (Sternberg, 2000). These aggravating factors suggest that "[i]ntelligence, as a coherent concept amenable to formal analysis, measurement, and duplication, may simply be an illusion" (Kaplan, 2016, p. 7). This is particularly problematic for its viability as a property capable of assisting with moral or legal judgments.

Dignity is the last of the ontological properties examined here, and the only one deserving of an entirely separate category as it relates to personhood, status, and rights. For reasons specified here, it might represent the hardest case for an attribute's translation to nonhumans. Kateb (2011) explains that human dignity comprises two essential claims: "[a]ll individuals are equal; no other species is equal to humanity" (p. 6). These ideas are reflected in foundational instruments of international human rights law. For instance, the very first provision of the Universal Declaration on Human Rights (UDHR) states unequivocally that "[a]ll human beings are born free and equal in dignity and rights" (UN General Assembly, 1948, art. 1). Metaphorically speaking, dignity is like "a special coin that is handed out to each person at birth" denoting his/her unique stature in the world relative to that of all other entities present (Daly, 2012, p. 14). This special quality reserved for humans alone emerges from our "rational nature" (P. Lee & George, 2008, p. 173), "practical autonomy" (Wise, 2002, p. 34), or "rational autonomy" (Tasioulas, 2019, p. 64). The possession or absence of dignity influences determinations of moral status (Shelton, 2014, pp. 7–8) and dignity rights (Daly, 2012, p. 6).

Animal rights theorists and practitioners have addressed the issue of dignity, albeit without the same fervor as they have other properties such as consciousness and sentience. Singer (1974) takes umbrage at the prized place dignity occupies in human ethical systems. He argues that dignity functions as a convenient, if ethically problematic, fast track to achieving an egalitarian society. After all, how else would one defend the conclusion that Adolf Hitler possesses more inherent worth than an elephant? As mentioned above, Wise (2013) finds that judges mainly view autonomy as the basis for dignity, so nonhuman creatures that exhibit autonomy should similarly be afforded dignity, which would qualify them for legal personhood and legal rights.

Not surprisingly, scholars writing on AI ethics have been even quieter on the subject of dignity. For instance, a global survey of 84 AI ethics guidelines finds that "dignity" was included in only 13 documents, making it the second-least common principle mentioned, ahead of only "solidarity" (Jobin et al., 2019). Only when the rights of humans are implicated in the development of artificial agents does dignity seem to enter the discussion (i.e., Donahoe & Metzger, 2019; Risse, 2019). For example, an analysis of robots applications in the healthcare domain suggests that human dignity should serve as the basis for governance over robotics (Zardiashvili & Fosch-Villaronga, 2020). However, when the issue of the dignity of intelligent machines themselves comes to the fore, scant mention of the property can be found. Notable exceptions include brief statements regarding the dignity of AI slaves (Solum, 1992, p. 1279), robots under the Japanese concept of *Mottainai* (Vallverdú, 2011, p. 181), and "hosts" (i.e., androids) in the HBO science fiction series *Westworld* (DiPaolo, 2019, p. 5). Like other properties before it, "dignity does not have a concrete meaning or a consistent definition" (M. Y. K. Lee, 2010, p. 157). In addition, its underlying reasoning is circular and appears to only reinforce a human-centered view of the world. Therefore, of the properties entertained here, dignity is perhaps the least likely to curry favor among those seeking to advance the rights of robots.

Societal need is the first of the relational mechanisms described in this section. Societal need is relational in the sense that it involves jurists acknowledging and responding to the imperatives of the wider community. The premise of this approach to determining legal personhood and legal status is simple—we should alter our legal constructs if doing so would help advance societal objectives. American jurisprudence has a longstanding tradition of applying this reasoning to corporations and ships. To be sure, no one seriously considers either of these nonhuman entities real, natural persons. Neither of them can reciprocate duties like true moral agents can (Watson, 1979, p. 123). But these concerns are beside the point. Treating companies and seafaring vessels as artificial persons helps to shield humans from liability and resolve conflicts. Thus, legal personhood has been extended to these entities because of the practical benefits obtained by treating them as more than mere things.

In Chapters Three and Four, I highlighted cases in which jurists performed nearly the same maneuver, although these disputes dealt with animals and nature. In *Karnail Singh*, an Indian High Court held that it was necessary to give rights to the entire animal kingdom in order to combat the scourge of environmental destruction. In *Chucho*, Judge Tolosa granted a writ of *habeas corpus* to release a bear from captivity as part of a larger effort to promote the survival of all species. Finally, in both *Ganges and Yamuna* and *Glaciers*, rivers were reconceived as legal persons as a means of safeguarding natural resources of great significance to the Hindu faith and community well-being.

Despite the problem-solving appeal of this jurisprudential innovation, there are at least two problems with promoting societal need as the basis for legal consideration. First, like environmental law in general (Boyd, 2003, p. 212; Kotzé & Kim, 2019, p. 5), experience has shown this strategy to be reactive, not proactive. Only after the lives of animals or the quality of the environment were threatened did we witness judges engage in the sort of conceptual broadening that expanded the realm of legal personhood. Until other countries follow suit, they are likely to suffer harms first and provide legal solutions later. Second, it is vulnerable to the criticism that this mechanism is both subjective and anti-democratic. Relying on unelected judges to act as "norm entrepreneurs" (Finnemore & Sikkink, 1998, p. 893) who reshape legal constructs in the face of societal challenges may lead to instability in the way legal systems interpret important terms and invite charges that legal actors are usurping the will of the people, invoking the specter of a "countermajoritarian difficulty" (Friedman, 2002, p. 155). All told, societal need offers much in the way of ideational flexibility and social responsiveness, but perhaps at the risk of institutional stability and democratic legitimacy.

Anthropomorphism is perhaps the most widely discussed mechanism by which humans relate to nonhuman entities. Despite being defined differently across disciplines, anthropomorphism is fairly well captured as "the tendency to attribute human characteristics to inanimate objects, animals and others with a view to helping us rationalise their actions" (Duffy, 2003, p. 180). In the context of moral and legal obligation, anthropomorphism leads to recognizing and privileging those entities that look and act like humans.[4] This tendency is nearly universal;

it can be found in numerous cultures across space and time (Boyer, 1996). In fact, anthropomorphizing other creatures may have endowed humans with an evolutionary advantage by helping hunters predict the moves of their prey, augmenting our cognitive capacity in the process (Mithen, 1996). The evolutionary antecedents of anthropomorphism continue to echo in modern times. For instance, a study by Miralles et al. (2019) suggests that humans feel more empathy and compassion towards species biologically closer to our own. Recent work from cultural psychology adds that perhaps social robots "exploit and feed upon processes and mechanisms that evolved for purposes that were originally completely alien to human–computer interactions" (Sætra, 2020). Today there is considerable debate over whether anthropomorphism is facilitated more by characteristics of the human perceiver or qualities of the object being perceived. Ample empirical research offers support for both of these arguments.

One explanation for variations in anthropomorphism focuses on the differences observed among humans. That is, the degree to which humans anthropomorphize other entities depends on their individual idiosyncrasies. Epley et al. (2007) argue that three psychological factors can affect the likelihood that humans engage in anthropomorphism: knowledge about humans and oneself (elicited agent knowledge), the desire to explain the behavior of other entities (effectance motivation), and the need for social interaction (sociability motivation). They find empirical support for effectance motivation and sociability motivation in experimental conditions involving animals (Epley et al., 2008). The same group of researchers also demonstrate that some people are more likely than others to anthropomorphize nonhuman agents (Waytz et al., 2010). In particular, they show that individuals who score higher in anthropomorphism (as assessed by responses to a questionnaire) are more likely to treat other entities as moral agents worthy of care and concern, believe that nonhuman agents are capable of intentional action and being held responsible for their behavior, and engage in socially desirable behavior when in the presence of such entities. Another analyst goes as far as to say that the attribution of human-like qualities is wholly independent from the nature of the object, and that anthropomorphism is instead stimulated by "relatedness, the disposition to consider others as possible interlocutors in communicative interactions" (Airenti, 2015, p. 123; emphasis omitted). When humans engage in relations with other agents on the basis that communication seems possible, empathy emerges.

Another explanation seeking to resolve differences in the extent to which humans anthropomorphize nonhuman agents centers on the traits of objects themselves. This angle helps us understand why "[p]eople ... anthropomorphize robots more than other technologies" (Young et al., 2011, p. 54). More specifically, anthropomorphism might vary according to an entity's physical appearance, movement, or behavior. Fifty years ago, Mori (1970) introduced the "uncanny valley" (*bukimi no tani*) hypothesis, which proposes a non-linear relationship between human-like appearance and affinity. The more similar an entity seems to a human being, the more positive affect it is likely to generate, but only to a point. At a certain level of human likeness, affinity plummets into an abyss of emotional repulsion. Affinity recovers and grows as human likeness escapes the valley, reaching its apex in the presence of an actual healthy human. In addition, Mori surmised that movement might dramatize the effect that appearance has on affinity.

This phenomenon has been explored extensively in the literature on human–robot interaction (HRI). While studies (con)testing the uncanny valley hypothesis are legion,[5] a few examples from HRI research may suffice to illustrate the scope of findings on the relationship between a robot's physical appearance and human affinity. One study concluded that human-like robots may be more likely than merely machine-like sophisticated robots to elicit changes in the non-verbal behavior of humans (Kanda et al., 2008). In a separate experiment, human subjects displayed more empathy towards a physically embodied robot than a disembodied one (Kwak et al., 2013). Researchers in another study determined that the morphology of a robot (i.e., whether it looks like a machine or a human) affected the likelihood that people assigned blame to the entity when it made a decision regarding a moral dilemma (Malle & Scheutz, 2016). There is also some evidence that movement might be even more important than appearance in fostering positive feelings about robots (Castro-González et al., 2016).

In addition to physical appearance, a robot's perceived behavior might also affect the likelihood that it garners affinity. Roboticists researching anthropomorphism have looked to the other side of the coin—dehumanization—in order to understand the conditions under which human similarity generates positive affect. On this subject, Haslam (2006) has identified two types of humanness—human uniqueness and human nature. Human uniqueness refers to qualities associated with civility, maturity, moral sensibility, rationality, and refinement. The absence of these traits may make an entity seem amoral, childlike, coarse, irrational, and uncivilized (i.e., animal-like behavior). Human nature, on the other hand, reflects agency, cognitive openness, depth, emotional responsiveness, and interpersonal warmth. Deficits in these areas are correlated with coldness, inertness, passivity, rigidity, and superficiality (i.e., machine-like behavior). Applying this typology, researchers have found that emotionality, not intelligence, makes robots seem more human-like (Złotowski et al., 2014), and that physical embodiment affects the perception of an entity's human nature, but not human uniqueness (Złotowski et al., 2015).

Other factors less well investigated than physical appearance, movement, and behavior are perceptions of mind and the context in which human–robot interactions take place. Gray and Wegner (2012) experimentally test the extent to which reactions associated with the uncanny valley hypothesis relate more to the perceived experience or agency of a robot and find support for the former. Fussell et al. (2008) consider the effects of repeated interactions with intelligent machines and observe that "as more people interact with robots, their abstract conceptions of them will become more anthropomorphic" (p. 145). Heeding the call to take context more seriously, Young et al. (2011) suggest analyzing HRI through the lens of "holistic interaction experience," which includes "visceral factors of interaction, social mechanics, and social structures" (p. 53).

Anthropomorphism offers a promising, but not bulletproof, mechanism for assessing the extent to which robots might be worthy of moral or legal concern. On the one hand, the tendency to assign human-like qualities to nonhuman agents is almost universally observed across cultures, is well researched, and offers an empirical basis for consequential determinations of personhood, status, and rights. On the other hand, there is a lot of conflicting evidence in the scholarly literature, variable levels of anthropomorphism among humans detract from the ability to make moral or legal determinations on a consistent basis, and the anthropocentric nature of the phenomenon might result in the exclusion or marginalization of less human-like entities. In sum, as a relational mechanism, anthropomorphism seems better equipped than most ontological properties to guide judgments about moral or legal consideration, but its case is far from airtight.

Kinship is the final relational mechanism evaluated here. Sahlins (2011) describes kinship as "'mutuality of being': people who are intrinsic to one another's existence" (p. 2). This concept involves a "compound and complex network of ties" (Malinowski, 1930, p. 29) that extends from the individual to the family to the clan and beyond. Fowler (2018) argues that both kinship—mutuality of being—and personhood—a state of being maintained through relationships—are useful for understanding social affiliations. In the context of moral or legal concern, kinship helps to specify the scope of the moral universe, the relations humans have with nonhumans, and the rules governing interactions between them. While the precise contours of these areas vary across traditional and Indigenous cultures, important similarities can also be observed.

Several different forms of kinship relations have been emphasized throughout this book. The cosmology of Cree people accommodates both animate and inanimate beings in a "circle of kinship" (Lewis et al., 2018, p. 7). The Lakota worldview remains open to the possibility that nonhuman, non-living entities can possess properties such as consciousness by virtue of the spirits that reside within them and the raw materials used in their construction (Posthumus, 2017). Adherents of Shintoism believe that both objects and natural beings have spirits, and that gods, humans, and nature are bound together through kinship because they are all related (Vallverdú, 2011). The Māori of New Zealand see animals, land, and plants as "imbued with spirits" that "also inhabit the people and look after them," resulting in reciprocal obligations among the community and the environment to nurture each other (Magallanes, 2015, p. 280). Despite the apparent diversity of ontological and cosmological perspectives identified above, a couple of unifying themes seem to emerge. First, kinship relations between humans and nonhumans are common among Indigenous societies (Studley, 2019). Second, many Indigenous groups espouse an understanding of the relationship between humans and the environment that might be characterized as a "kincentric ecology" (Salmón, 2000, p. 1327), in which humans are but one strand in a larger web of relations that fosters mutual responsibility among its constituent elements. Notably, these relationships need not exist in dyadic form exclusively among humans and nature. For instance, there is some scientific evidence indicating that trees communicate with each other and have familial relations of their own (Wohlleben, 2016).

As a criterion used to determine moral or legal obligation, kinship holds great potential. Although it may mean somewhat different things in different social contexts, common threads such as recognition of nonhuman entities (even AI) and ecological sensitivity provide some room for cross-cultural application. Further, kinship, more directly than any of the other properties or mechanisms reviewed in this section, abjures anthropocentrism, which has shown to be problematic for both ethical and environmental reasons. Finally, kinship deflates Western chauvinism by acknowledging, celebrating, and elevating the worldviews of Indigenous and traditional peoples. These positive attributes do not render kinship immune to criticism, however. For one, the perspectives found among non-Western societies regarding social relations may present practical problems where legal systems are not coterminous with cultural geographies. More bluntly, how will we know what standard to use in a given jurisdiction? In addition, the diversity of views about kinship may lead to confusion and/or inconsistent outcomes. Lastly, those people (i.e., non-Indigenous persons) charged with representing or adjudicating the interests of nonhumans may default to anthropocentric behavior given that the responsibility for making moral or legal judgments ultimately lies within the hands of human actors. In other words, implementing kinship might prove difficult for people alienated from or unfamiliar with kincentric ecological cosmologies.

This section explored the viability of a number of ontological properties and relational mechanisms for determining the moral or legal status of nonhuman entities. By no means was this exercise intended to be exhaustive. Several more properties and mechanisms have been mentioned throughout this book, including, *inter alia*, rationality, will, capacity to be a legal subject, animism, facing, and self-realization. My goal in this part of the final chapter has been to assess the individual candidacy of arguably the most common and controversial characteristics or ways of relating to others in an attempt to locate suitable candidates for a uni-criterial approach to moral or legal consideration. This attempt has proven unsuccessful, and that is precisely the point. None of the properties or mechanisms alleged to inform personhood, status, or rights is independently sufficient for such purposes. All of the criteria examined above feature strengths and weaknesses. In general, while properties provide parsimony, relations revel in complexity. Yet, as I have shown, relations are more inclusive of various entities than are properties. To reiterate a point made in Chapter Two, properties and relations are inextricably linked; the latter are often informed by the former. Therefore, it is an inescapable fact that both properties and relations will necessarily invade analyses of moral and legal obligation. The next two sections seek to build on the conclusions reached in this section by extracting insights from the theory and practice on animal rights and the rights of nature.

Animal rights: Revealing tensions, not solutions

The literature on animal rights reveals four main tensions left unresolved in the quest to extend enhanced moral or legal protection to nonhuman entities. First, at least some degree of anthropocentrism seems unavoidable, exposing animal rights to the criticism that humans will always enjoy a status superior to that of all other entities. Second, the importance of reciprocal duties remains unsettled, with attendant implications for the likelihood that moral patients like animals (or perhaps robots) might be deemed worthy of moral rights. Third, definitional, conceptual, and empirical issues associated with various ontological properties (i.e., consciousness and sentience) render their applicability to moral and legal analyses suspect, as evidenced by the argument from marginal cases. Fourth, and on a related note, moral and legal theorists alike have struggled to advance a logically defensible and inclusive program for identifying those animals eligible for personhood, status, and rights. These flaws might be overcome if any or all of the following were achieved: (1) scholars and jurists agree on the relative importance of certain properties; (2) there was a compelling societal need to extend rights to a specific animal or group of animals; and (3) Western philosophy and law became more open to adopting insights from non-Western worldviews.

Case law on animal rights appears to offer more promise than animal rights theory. Litigation has been initiated by humans acting on behalf of animals on the basis of alleged harms to the animals themselves, not humans or their sensibilities. Judges (mainly those in the developing world) have exhibited a willingness to consult a breathtaking array of foreign and domestic laws, scholarship, and cultural practices in their deliberations. Decisions finding that animals indeed possess rights have also relied on a mix of approaches as opposed to remaining content to adjudicate claims using a single analytical lens. Finally, animal rights cases suggest that how we choose to define terms integral to judicial decision-making where nonhuman entities are concerned (i.e., legal personhood, legal subject, and legal rights) greatly affects the kinds of outcomes obtained.

In sum, while theory on animal rights is tangled and irresolute, the associated jurisprudence has proven flexible and innovative. The properties-based approach still tends to dominate discussions about the moral and legal status of nonhumans, although relational approaches have made modest inroads in academic circles and legal decisions. However, scholars and practitioners have yet to devise a coherent and consistent model capable of being deployed in different political and legal contexts. The lingering issues enumerated above will need to be tended to if animal rights are to take flight on a global scale and prove useful in the domain of intelligent machines.

Rights of nature: Embracing vulnerability and contingency in the Anthropocene

Scholarship relevant to the RoN provides fertile ground for cultivating moral and legal concern regarding natural and artefactual nonhuman entities. The environmental ethic of ecocentrism, but more specifically its cognate progeny of deep ecology and transpersonal ecology, delivers an ontological orientation (i.e., holism) that is inclusive of a range of biotic and abiotic forms and resonates with non-Western perspectives found in Buddhism and Indigenous cultures. Critical environmental law infuses the discussion with radical departures from con-ventional Western thinking by replacing Cartesian dualisms with assemblages, de-centering humans from their presumed position of authority, eschewing axi-ological and utilitarian logics in favor of recognizing a vulnerability common to all, decolonizing environmental law, and embracing the contingency and fluidity inherent to an open ecology. Writing on environmental law in the Anthropocene instructs that this geological epoch presents an opportunity for assessing the fail-ures of anthropocentric Western institutions and considering bold alternate imagi-naries that disrupt the path dependent tendencies hastening planetary destruction. New Materialism adds the insight that all matter—organic and inorganic—is endowed with agentic capacity, annulling the idea that humans are the only agents in the moral universe.

Cases involving the RoN illustrate how holistic environmental ethics and radi-cal reform programs might translate in different legal contexts. Causes of action regarding alleged harm to humans and nature have been adjudicated through the lens of ecocentrism, although the complaints have focused on redressing past or ongoing grievances instead of anticipating them. To reach their decisions, jurists have sought the counsel of numerous sources, including foreign and domestic law, Indigenous ideas, and religious beliefs. Critical environmental law has also crept into RoN jurisprudence, as evidenced by legal reasoning that de-centers humans, de-individualizes legal subjects, or expands the definition of a legal per-son beyond its standard contours. Interestingly, the outcomes observed suggest that radical decisions invoking global norms might still be filtered through local contexts, as Latin American cases expanded the purview of legal concern to non-human *natural* entities while Indian cases broadened the ontological horizon even further by recognizing the legal personhood of non-living *artefactual* entities.

Scholarship on environmental ethics and environmental law obtained concrete (and occasionally explicit) expression in case law dealing with the RoN. In the course of this harmonizing, hitherto unrecognized nonhuman entities emerged as new legal subjects in an open ecology without the requisite agonizing over the extent to which they possess morally significant properties. As such, RoN jurisprudence has contemplated the legal status of nature using methods totally alien to the doctrinaire rubrics employed in animal rights literature and litigation. However, the reasoning that facilitated this ontological shift has shown itself to be reactive, eclectic, and context-specific, hardly a portable recipe for combatting environmental destruction in disparate corners of the Earth. It remains to be seen whether jurists are willing to apply a similar approach to questions with arguably less existential urgency, like the legal personhood of technological beings.

Towards a multi-spectral framework for determining personhoods

In order to assess the extent to which nonhuman entities might be at least theoreti-cally eligible for moral or legal consideration, three questions must be addressed. First, *what is the nature of the inquiry—moral or legal*? As I have suggested throughout this book, the latter is easier to resolve than the former. Moral consideration has proven seriously constrained by the ambiguity surrounding the selection of morally relevant properties. Barring the arrival of paradigm-altering empirical revelations about consciousness, intelligence, sentience, and so on, the properties-based approach to moral status will remain mired in a vigorous, inconclusive debate among philosophers. A more productive way forward has been proposed by Levinasian scholars, namely Coeckelbergh, Crowe, and Gunkel, who have injected new life into this old discussion by reversing the Humean thesis to elevate the nature of encounters with alterity above properties alleged to provide a means of morally discriminating among entities. However, this relational approach tends to downplay the important role that properties play *during* these encounters.

In terms of legal issues, Chapters Two and Four, along with the previous section of the present chapter, demonstrate that legal personhood, legal status, and/or legal rights have been extended to plenty of nonhuman entities in both developed and developing countries for reasons related to expediency and societal need. Such legal persons have included bears, chimpanzees, corporations, idols, nature, rivers, ships, and even the entire animal kingdom. Conversely, nonhumans have been denied legal benefits where jurists have engaged in the rigid application of legal definitions, exhibited antipathy towards foreign legal ideas and non-Western worldviews, and tread upon the messy terrain of criteria relevant to moral obligation, an exercise that is unnecessary for matters regarding legal status. To be direct, there is nothing prohibiting nonhumans from obtaining legal recognition of one kind or another provided that doing so is considered necessary to advance societal goals or overcome societal problems. Thus far, judges have kept careful watch over the ontological boundaries of the legal realm, only occasionally expanding preconceived notions about who counts as a person in a given jurisdiction.

Second, *does the ability of an entity to reciprocate duties or be held responsible matter for the assignment of moral or legal rights?* Moral rights have long been reserved exclusively for moral agents capable of fulfilling moral duties given their possession of certain ontological properties. However, there are some humans who fail to pass the test of moral agency (i.e., fetuses, infants, the mentally incapacitated). In these marginal cases, the possession of moral rights is determined by their dignity and inherent worth as humans. This logic is incomplete and circular, rendering its conclusions doubtful. In addition, as detailed in this chapter and in Chapter Four, New Materialism makes a compelling argument that agency (along with some of its cognate properties) is "distributed across a vast range of beings and entities" and "decoupled from humanity" (Arias-Maldonado, 2019, p. 53). An alternative and complementary concept that attempts to fix the flaws of moral agency is moral patiency. Unlike moral agents, moral patients do not suffer the burden of enacting duties. They are moral patients by virtue of the fact that things done to them can be considered right or wrong. For some, any bit of information could qualify as a moral patient (i.e., Floridi, 1999). For others, only those entities who are owed a duty by a moral agent can be considered for this status (i.e., Himma, 2009). Yet even this new category of moral entity is not without its share of practical and epistemological problems (Gunkel, 2012, pp. 153–157). Still, the ongoing project of refining moral patiency sheds light on the limited analytical usefulness of relying on moral agency as the sole benchmark for the extension of moral rights.

On the legal side, Western systems of law have long separated legal subjects worthy of rights from legal objects that amount to property. But the content of these categories has been reshuffled throughout history, perhaps most notably in the case of slavery. The fact that some humans were once regarded as property to whom no legal duties were owed exposes legal status as historically contingent and conceptually malleable. As the category of legal subjects has become more inclusive and humanity has sought to resolve more of its disputes in the court-room, some jurisdictions have taken up the mantle of expanding notions of legal personhood and legal status, affording ever more entities the opportunity to obtain legal recognition. An example of this kind of progressive legal decision-making can be seen in the *Glaciers* case reviewed in Chapter Four, in which nature was granted legal rights on the basis of moral duties humans have towards future gen-erations. Thus, the capacity to reciprocate legal duties is not a condition necessary for obtaining elevated legal status or legal rights.

Third, *what role do properties and/or relations play?* As stated numerous times throughout this text, using ontological properties as the basis for psycho-logical/moral personhood, moral status, and moral rights is fraught with compli-cations. Importantly, substantial disagreement persists regarding the quality(ies) deemed necessary for moral personhood and how the presence of one or more of these traits in an entity can be empirically verified. Legal personhood has also succumbed to the allure of properties in two ways. First, the capacity to be a legal subject, possess rights, or fulfill duties is determined by the possession of cer-tain properties. Second, even the concept of dignity derives its significance from autonomy, a property associated with moral personhood. Thus, where moral or legal obligations towards nonhuman entities are concerned, some combination of ontological properties is required. Relations, by contrast, are more complex and contingent, contributing to relational personhood, which may be seen as occur-ring prior to judgments regarding an entity's eligibility for psychological/moral or legal personhoods. Yet, even relational approaches do not fully extricate proper-ties from assessments of moral or legal concern. This is because our encounters with alterity are colored by human tendencies—like anthropomorphism—which privilege certain forms and attributes above others. Although humans possess sub-conscious preferences for specific features, these physical, cognitive, and emo-tional qualities collectively assume neither the shape of a generalizable archetype nor the form of a near-perfect human facsimile, as evidenced by HRI research on the uncanny valley. All of this is to say that despite the problems identified with properties, the approach advocated here—an explicitly relational one—must nec-essarily incorporate such perceived characteristics to some extent.

So far we have determined that (1) deciphering an entity's eligibility for rights depends on whether the inquiry is focused on either the moral or legal dimen-sion; (2) agency is flawed as a standard for moral or legal status and the ability to reciprocate duties is not required for either moral or legal consideration; and (3) a relational approach will have to account for properties as well. The next step towards designing a framework that helps determine the kind(s) of personhood for which an entity might be eligible requires two additional specifications.

First, efforts to assess an entity's candidacy for moral or legal concern must transition from being uni-criterial (i.e., Singer's (1974) sentience) or multi-criterial (i.e., Warren's (1997) properties and relations) to multi-*spectral* (i.e., Fowler's (2018) dimensions of personhood). Parsimony should give way to complexity, allowing for variability in the importance and intensity of certain factors used in the assessment of personhood. Second, the philosophical approach undergirding both moral and legal concern should consist of an amalgamation of properties and relations clumsily but accurately described as "properties-as-they-appear-to-us within a social-relational, social-ecological context" (Coeckelbergh, 2010, p. 219). There are five important elements in this long-winded phrase—*properties*, *us*, *relational*, *ecological*, and *social*. *Properties* possess an undefined but undeniable currency that generates affective and behavioral responses among those entities present on a common surface. *Us* should be interpreted in a holistic and non-anthropocentric manner, meaning something closer to "whomever or whatever." Latour's concepts of "actant" and "hybrid" offer some analytical clarity here. Actants refer to humans and nonhumans, categories of being that replace subjects and objects (Latour, 2004, p. 237). Hybrids consist of "associations of human actors and nonhuman actants" (Teubner, 2006, p. 511). In terms of the model proposed here, entities present during social encounters are either actants or members of a hybrid engaging in relations with other beings under a broader ecological context. As such, *us* could describe humans, bears, nature, idols, robots, or any combination thereof. *Relational* and *ecological* refer to the scope and content of the contexts in which encounters occur. The prefix *social* entails how identities are co-constructed through mutual relations.

The framework presented here builds off the conceptual map from Chapter Two in the form of a multi-spectral tool that draws together all four types of personhood and their requisite ontological properties or relational mechanisms, each of which may be more or less significant in a given context (see Figure 5.1).

As indicated in the figure, there are six criterial categories that correspond to the four kinds of personhood. *Mentality* includes consciousness, intentionality, and sentience. *Autonomy* consists of intelligence, rationality, responsibility, and will. *Capacity* refers to the ability to hold rights, burdens, and entitlements. *Utility* reflects the practical reasons for extending legal personhood. *Contextuality* entails the extent to which an assemblage is embedded in the social world and ecological system, and the relationship between bodies in the assemblage. Finally, *physicality* involves the material composition of the entities (i.e., partial determinants of zoo- or anthropo-morphism) and their appearance-in-context to actants or hybrids.[6] Importantly, there are no minimum necessary conditions or thresholds beyond which an entity can be said to possess one or another kind of personhood (as evidenced by the plus and minus signs). Each category is ontologically flexible, epistemologically contingent, and context dependent. Just like in the conceptual map presented in Chapter Two,

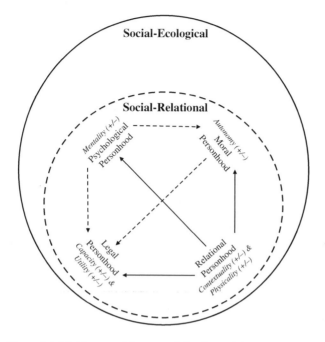

Figure 5.1 Multi-spectral framework for determining personhoods.

arrows connecting the personhoods reflect the directionality of influence between them and solid lines reflect stronger associations than dotted ones. Each of the forms of personhood are nested within social-relational and then social-ecological contexts, which interact with one another, as evidenced by the dashed line. Using this framework, determinations of personhood(s) can be made using a method similar to the one employed by the judges in *Karnail Singh*, which relied on properties, relations, laws, jurisprudence, societal need, and religious doctrine.

Towards a critical environmental ethic

The second move in the project of incorporating artefactual entities into moral or legal universes involves proposing an ethical outlook hospitable to their inclusion. In the space here, I will crudely sketch the contours of what might be called a *critical* environmental ethic conceived with this purpose in mind.

First, the ontological orientation should shun individualism in favor of a radical, not merely biospherical, holism. Given the collapse of the human–nature Cartesian divide, it is no longer appropriate to talk of nature as separate from human society or entities as existing wholly separate from each other. As Vogel (2015) boldly asserts, "[t]he distinction between the natural and the artificial is ontologically meaningless" (p. 169). If there is no ontological difference between nature and culture, artefacticity becomes a relic of bygone binaries. Robots, previously considered cultural artefacts "discontinuous with nature" (Bryson, 2018, p. 17), are now woven into a "plane of immanence" in which differences recede and the defunct categories of artificial, human, natural, and technological "fold into each other and constantly emerge as epistemological and ontological hybrids" (Philippopoulos-Mihalopoulos, 2013, p. 857). What surfaces is a "singularity" that is a feature of Anthropocenic radical acentricity (Philippopoulos-Mihalopoulos, 2017, p. 133), not the "technological singularity" of AI lore (Winfield, 2014).

Second, and on a related note, anthropocentrism should be rejected in favor of an ecocentric-anthropic ethical perspective. This ethical position, borrowed from deep ecology and Judge Tolosa's decision in *Chucho*, adopts a "relational, total-field image" (Naess, 1973, p. 95; emphasis omitted) that nevertheless recognizes the important obligation humans have to protect the whole environment and all its inhabitants, and their unique capacity to realize a more benevolent world.[7] Of course, this perspective presents a paradox. That is, in the same breath it advocates in favor of refocusing the ethical universe through ecocentrism while simultaneously reifying nature as a subject. However, a critical environmental ethic acknowledges rather than dismisses or retreats in the face of incongruences, contradictions, complexities, and slippages inherent to the Anthropocene (De Lucia, 2017, p. 116).

Third, the kinds of subjects that fall within the ethical domain are all those that find themselves vulnerable in an "'open ecology' of social, biological and ecological processes" (Philippopoulos-Mihalopoulos, 2013, p. 854). These include assemblages of hitherto discrete beings that are in actuality inextricably bound together. A couple of concrete examples may suffice to illustrate this idea. In *Chucho*, Judge Tolosa gave legal effect to the term "nature–subject couplet" (emphasis omitted), which describes the union of a legal subject and the environment in which it exists.[8] In the context of modern warfare, Arvidsson (2020) describes AI swarming insect drones as indicative of the "posthuman condition of an ecology of human–animal–technology entanglement" (p. 135). In both examples, it does not make sense to extract individuals from these assemblages, as their existences are thoroughly and indissolubly fused. Furthermore, such groupings may expand or contract depending on the context, reflecting "an ontology of continuous connection between bodies" (Philippopoulos-Mihalopoulos, 2017, p. 132). These "bodies melting into each other's contours" constitute an "acentral and multi-agentic" continuum (Philippopoulos-Mihalopoulos, 2017, p. 123). This critical awareness of instability, contingency, and vulnerability materializes as one of the chief insights afforded by the Anthropocene, or what Haraway (2016) refers to as the "Chthulucene," which "is made up of ongoing multispecies stories and practices of becoming-with" (p. 55).

Fourth, the Anthropocene (or Chthulucene) provides the moment of crisis and opportunity for revision that enables the elevation of new cosmologies and imaginaries, like those of traditional and Indigenous cultures. Under this new state of affairs, it is no longer permissible to exclude from the realm of possibility other "ways of worlding" (Blaser, 2014, p. 54). As such, Western moral and legal systems are forced onto a plane of radical equality with non-Western perspectives that may introduce concepts unfamiliar to or even uncomfortably challenging their existing modes of operation. Examples of this critical approach in action can be seen to some extent in all four RoN cases discussed in Chapter Four, in which nonhuman entities were recognized as legal persons with legal rights. Paying tribute to Indigenous cosmologies, this ethical perspective is as kincentric as it is ecocentric-anthropic.

Fifth, the content of this ethical orientation demands a heightened awareness of and responsiveness to several concerns (Grear, 2015b, pp. 304–306). Such an ethic must first and foremost be ecological in terms of both its normative objective and epistemological sensibility. That is, it should seek to improve and protect the environment while remaining sensitive to the insights obtained through various relations, behaviors, and knowledges. This means honoring the actual lived experiences and condition of all entities, assemblages, and systems. In addition, substantial emphasis should be placed on combatting inequality and injustice in all forms, which requires acknowledging the disproportionate violence dealt to marginalized human populations and nonhuman animals (and perhaps one day artificial agents, if it is not already the case) and "deliver[ing] … inclusion, compassion and resilience" (Grear, 2015a, p. 246). This ethic also recognizes the power associated with subject-positionality and its manifestation in uneven, particular, and variably experienced vulnerabilities. Finally, a critical environmental ethic pays special tribute to the intrinsic linkages between macro-level phenomena and micro-level concerns. As Dr. Martin Luther King, Jr. (1963) once wrote, "[i]njustice anywhere is a threat to justice everywhere" (p. 2).

Sixth, rights hold a contradictory status in a critical environmental ethic. On the one hand, rights represent the failed anthropocentric instrument of a predominantly Western legal order. On the other hand, the Anthropocene creates a fissure whereby alternative interpretations of rights are given the space to blossom. It is therefore perhaps best to think of rights under this ethical approach as delivering a range of options that vary in terms of their transformative potential. Arguably the most transformative (but also the least innovative) are constitutional environmental rights (Collins, 2019, p. 13). Other avenues include rights from Indigenous law and ecological law. Indigenous legal systems have dealt with environmental concerns for far longer than industrialized countries have, and they are far more accommodating of nonhuman entities. Ecological law, which is based on ecocentrism, seeks to spread an ecological consciousness throughout all legal and political institutions. A critical environmental ethic would remain open to any of these foundations for rights. Importantly, this openness could support the extension of rights to any entities or constituents of assemblages given the ethic's rejection of dualisms and hierarchies that have justified longstanding ethical intolerance towards nonhumans.

To briefly restate, a critical environmental ethic is holist and ecocentric-anthropic; includes all vulnerable bodies present in an open ecology as assemblages; accommodates non-Western worldviews (especially the notion of kincentric ecology); requires responsiveness to ecological concerns, lived experiences, inequality, power, positionality, vulnerability, and multi-level inter-connections; and is open to extending rights to nonhuman entities.

Applying the multi-spectral framework and critical environmental ethic

In order to evaluate the extent to which a nonhuman entity such as an intelligent machine might be eligible for rights using the framework and ethic proposed here, several questions must be asked (although not necessarily in the following order). First, what kind of personhood is under investigation—psychological/moral or legal? The answer to this question will guide the observer down pathways to different forms of personhood, status, and rights, as discussed earlier. Second, what are the properties of the entity in question as they appear to relevant actants or hybrids in the context of the social interaction and broader ecological context? Third, what is the relative importance of these properties in the specific relational and ecological contexts? Fourth, what is the nature of the relations among the actants and/or hybrids in terms of both the encounter and the environment? Fifth, what are the scopes of the social-relational and social-ecological contexts? Sixth, how do the aforementioned factors shape expectations about the kind of conduct permissible among the entities involved? To lend a critical bent to Leopold, *a thing is right when it tends to deliver inclusion, compassion, and resilience to all present in an open ecology. It is wrong when it tends otherwise.*

What might implementation of a multi-spectral, critical approach look like? To begin, we would have to choose whether to focus the analysis on the qualification for psychological/moral or legal personhood, realizing that the path pursued might wind up being epiphenomenal to relational personhood. The former would involve simultaneous consideration of both properties and relations. This entails demonstrating the degree to which the observer perceives the intelligent machine to possess some combination of attributes that support the act of identification (i.e., cosmological, personal, or ontological). Crucially, the mix of qualities deemed relevant will vary to some extent from actant to actant, and some may feature more prominently than others in the course of identifying (with) a robot. For instance, some human partners may privilege verisimilitude with respect to consciousness, while for others, approximating intelligence is more important. In addition, understanding the temporal, spatial, and phenomenological dimensions of the encounter will flesh out the nature of the contexts in which the interaction occurs. This interactive process of interpretation, considered in tandem with an assessment of the relational and ecological contexts in which the encounter takes place, determines the likelihood that an actant qualifies for relational personhood, which in turn affects the robot's candidacy for moral personhood. Significantly, in line with the relational approaches to robot ethics reviewed in Chapter One, the proper ethical orientation an actant should apply when encountering an intelligent machine is a function of their mutual relations. Appropriate behavior towards alterity is predicated not primarily on the properties they are perceived to have, but rather on the conjunction of traits perceived by the observer and the nature of the moment in which AI reveals itself to its reciprocating actant. In other words, the qualities of the entity structure but do not wholly determine the ethical orientation applied in the course of mutual engagement. Upon qualifying for moral personhood, further inquiry is made as to whether the robot might be considered either a moral agent or moral patient. If shown to possess some level of responsibility or accountability for its actions separate from that of its programmers or users, the intelligent machine may be viewed as a moral agent. If judged otherwise, the robot would likely hold the status of moral patient. The specific types of moral rights for which the AI is eligible would vary according to its moral status.

If legal personhood was of interest instead, the analysis would proceed differently. The multi-spectral framework suggests that personhood in the legal domain is determined by two factors—capacity and utility. Capacity is often tied to properties such as autonomy, consciousness, and rationality. Utility entails practical reasons why society might benefit from extending legal personhood to an entity, such as assisting in the resolution of disputes involving nonhumans. Importantly, assessments of both capacity and utility cannot be divorced from the relational and ecological contexts in which they are conducted. Pertinent issues along these lines include the nature, duration, and intensity of relationships among entities and/or the extent to which they comprise an assemblage; the situatedness of the parties (i.e., their positionality) within the larger ecological scheme; and the degree of kinship that an observer feels with an actant. In the event that a robot is found to possess sufficient capacity and/or utility relative to its relational and ecological contexts, it may qualify for legal personhood. Since a critical environmental ethic does not distinguish between legal objects and legal subjects, the next step would focus on the specific kinds of legal rights to which the AI might be entitled.

In terms of both moral and legal rights, the precise content of those entitlements would be shaped by the normative commitments held by a critical environmental ethic. That is, rights that serve to promote inclusion, compassion, and resilience, and eradicate injustice and inequality, would be extended to the appropriate actants and/or hybrids.

Robot rights on the horizon? Two hypothetical examples

In an effort to illustrate how the framework and ethic developed in this chapter might be used to determine the applicability of rights to intelligent machines, in this section I present two hypothetical examples drawn from robotic technology currently out in the world—zoomorphic robot companions and anthropomorphic sex robots.

Zoomorphic robot companions (i.e., robot dogs and cats) have been created with the intention of providing comfort and entertainment to owners (especially the elderly) without requiring the work normally associated with caring for live animals (Sparrow, 2002). Empirical evidence suggests that humans express uncertainty about the ontological status of these robot companions. An analysis of three studies pertaining to human interactions with robot dogs shows that people perceive these intelligent pets as "hybrid" entities—"technological artifact[s] that also embod[y] attributes of living animals, such as having mental states, being a social other, and having moral standing" (Melson et al., 2009, p. 546). This ambivalence has already produced some interesting real-world impacts. For instance, in 2019, YouTube's algorithm began removing robot combat videos on the grounds that they ran afoul of community standards prohibiting depictions of animal cruelty (Cuthbertson, 2019). Although the robots featured in these videos did not look much like quadrupedal domesticated animals, the controversy highlighted concerns about how humans identify and treat nonhumans, even those of the zoomorphic mechanical kind. On a more anecdotal basis, roboticist Atsuo Takanishi recounted during an interview how his mother would often talk to Aibo (a robotic dog) as if it were a "human baby" or "pet animal."

Applying a uni-criterial approach to animal rights in the context of zoomorphic robot companions offers only a simplistic analysis that would likely end in a most obvious conclusion—robot pets are neither sentient nor subjects-of-a-life; therefore, they are eligible for neither moral nor legal personhood. Considering a single relational mechanism like anthropomorphism would also not do much to resolve a robot pet's moral or legal status, as that tendency has proven highly variable among humans, and less is known about the translation of this phenomenon to intelligent zoomorphic machines. A multi-criterial approach (i.e., Warren, 1997) still leaves guesswork to be done as to the presence of certain morally significant characteristics, relies on a hierarchy that places properties above relations, neglects the role of moral patiency, defines artefacts narrowly, and reifies Cartesian boundaries cast into doubt by the Anthropocene. Therefore, as argued above, both uni-criterial and multi-criterial approaches remain deeply unsatisfactory.

A multi-spectral, critical approach provides a more complex method for conducting this inquiry. Consider a scenario in which a robot dog serves as a companion for an elderly person. Immediately upon purchasing her new robot dog, the owner gives it a name—"Fi-Do" (short for "Fidelity-to-Dog"). As the owner lives alone, Fi-Do provides an important source of daily interaction and comfort. Despite have lost much of her sight due to cataracts, the owner has managed to retain an acute sense of hearing. As such, the elderly woman speaks to Fi-Do constantly throughout the day, and her spirits are lifted when her prized pet barks back in response or cozies up to her while she's sitting on her couch. One day, while the owner is out buying groceries, a burglar breaks into her apartment, stealing valuable personal items, including Fi-Do's expensive charging station. Without the charging station, the elderly woman is unable to recharge her companion, rendering Fi-Do helplessly inert.

Does Fi-Do have rights, and, if so, were they violated by the burglar? The robot dog's eligibility for moral or legal personhood (and thus moral or legal rights) would be driven by the initial encounter with its owner, whose perception of and reaction to the mechanical pet would be informed by a combination of *a priori* expectations and conditions inherent to the relationship as it unfolds (i.e., contextuality). Naming the robot companion indicates personification, and the positive affect that the owner feels towards the intelligent machine demonstrates a kind of felt nearness, both of which suggest relational mechanisms at work. The owner, whose sight isn't what it used to be, may have a better sense of the robot dog's simulated sentience than its intentionality (i.e., psychological properties), because she can hear the dog yelp when it "accidentally" runs into a wall. The human–robot dog relationship would also be colored by the broader social-ecological context. For instance, the robot companion might be considered a member of the owner's household within a society that bestows special spiritual value upon dogs and technology, both for reasons related to non-Western religious beliefs. In addition, the robot dog's unique positionality (i.e., a companion to a visually impaired elderly woman) and vulnerability (i.e., it needs to be charged in order to continue operating, making it dependent upon others) would be taken into account when seeking to understand the ethical implications of actions that pose an affront to the hybrid entity's well-being (i.e., depriving the robot of its charging station). The robot dog's moral personhood would then be assessed based on these initial conditions, and its potential for obtaining legal personhood would be determined by analyzing the pet's capacity and utility. If, for instance, designating Fi-Do a legal person for the purpose of filing a claim might have the practical effect of deterring would-be burglars from robbing elderly people of their ability to interact with their robot companions or inspiring people to treat both robotic and live animals more humanely, these factors could be incorporated into the equation. Finally, any specific moral or legal rights Fi-Do might possess in light of the aforementioned factors could be explicitly set forth by legislators or judges.

Anthropomorphic (or humanoid) sex robots, at one time only the stuff of science fiction or speculative fascination, have also recently become tangible fixtures of modern society. Broadly speaking, these intelligent machines are "[c]reated specifically to allow individuals to simulate erotic and romantic experiences with a seemingly alive and present human being" (Gersen, 2019, p. 1793). There are two main schools of thought regarding the ethical and legal implications of sex robots. One group of observers presents arguments in favor of the design, sale, and use of sex robots. Some, like Ron Arkin, director of the Mobile Robot Laboratory at Georgia Tech, suggest that they might be useful for research purposes (Hill, 2014). Others have made the case that sex robots might help physically and mentally disabled people enjoy the right to sexual satisfaction (Nucci, 2017). Still others approaching the subject from queer studies propose that "sex robots may in fact enable new liberated forms of sexual pleasure beyond fixed normalizations" (Kubes, 2019, p. 1). Libertarians might contend that people should have the freedom to purchase any goods that the market will provide and government should not regulate provision of those goods (Danaher, 2019). Another group of scholars identifies major ethical or legal concerns regarding sex robots. These analysts raise a number of objections, including that sex robots may cause humans to ignore their "full humanity" (Richardson, 2016, p. 52), experience difficulty establishing meaningful relationships with other humans (Nyholm & Frank, 2019), and engage in practices that increase their propensity to harm other people (Sparrow, 2017).

A uni-criterial method of determining moral or legal obligations towards sex robots would likely focus on evidence suggesting the possession of autonomy, consciousness, or intelligence. While a survey of present technological capabilities might quickly dispense with the idea that any of these apply, it is at least conceivable that future iterations of sex robots might come close to approximating these qualities in human-like proportions. However, demonstration of any one of these properties might be insufficient to obtain moral or legal status due to the lingering issue of the argument from marginal cases. It strains credulity to think that humanity would settle on a particular trait that might not also be present (at least not to a high degree) in a flesh and blood human. Relying on a sole relational mechanism like societal need might also prove lackluster, as that might assist in determining legal, but not moral, status. A multi-criterial approach would afford the inquiry more room for acknowledging complexity, but it would also likely classify a sex robot as a kind of mere artefact deserving of less moral or legal status than other entities whose possession of certain properties grants them a position above that of other candidates. In addition, this type of approach would still maintain ontological separations that are no longer tenable in an era when the boundaries between humans, nature, and technology are blurrier than ever.

A multi-spectral, critical approach would be better suited to handle the unique vulnerability and contingent status of sex robots. The strength of this argument can be illustrated through the presentation of a hypothetical example. Imagine a lonely but otherwise neurotypical human who purchases a sex robot ("Pat") designed to provide its owner with a means of achieving sexual gratification. The human, who comes to feel love and affection for the mechanical mate, often brings the robot into public spaces to serve as a companion. Amorous activities between the two, however, occur exclusively in the confines of the owner's private residence. One day, acting out of self-pity due to an inability to perform sexually the previous evening, the owner strikes Pat across its face while the two ride the subway together, much to the horror of nearby passengers, some of whom confuse the robot for a human person until authorities arrive on the scene.

The likelihood that this intelligent machine might qualify for moral personhood would depend on several factors, including the extent to which both the owner and the onlookers perceive the robot to possess attributes associated with mentality and/or autonomy (i.e., consciousness or intelligence), the nature of the relational (i.e., character of human–hybrid interactions) and ecological (i.e., public transportation setting nested inside of the broader physical environment) contexts, and the material composition of the humanoid (i.e., silicone, steel, and thermoplastic elastomers processed with oil). Determining the potential that Pat might be eligible for legal personhood would require additional inquiries into the robot's capacity (i.e., evidence of properties intrinsic to psychological and moral personhoods) and utility (i.e., strengthening societal norms against domestic violence). Furthermore, the physicality of the sex robot might invoke Indigenous claims of kinship given the use of raw materials imbued with spirits that were used in the construction of the technological entity. As suggested here, both moral and legal personhood are fundamentally derived from relations between the robot and other actants present within its ontological orbit. Finally, the nature of the owner's conduct would also influence the task of assessing personhood, as the reprehensible display of power and positionality committed against a vulnerable nonhuman would be deemed unequivocally unjust and, therefore, unethical. The specific moral or legal rights that would apply to Pat in this situation could either be prescribed proactively through legislation or reactively through judicial decision-making.

Both of these hypothetical examples were provided in order to demonstrate how the multi-spectral framework and critical environmental ethic might be used to respond to real-world situations where the rights of intelligent machines are at issue. To be sure, the application of these devices is not likely to follow a linear path; they are menus, not quadratic equations. However, I hope that the approach suggested here proves useful for pushing the conversation about the extent of our obligations towards nonhumans in general and robots in particular in productive directions.

Looking forward

While this chapter has sought to develop a method for assessing the conditions under which robots might be eligible for rights, more work remains to be done on this and related inquiries. In this last section, I identify three promising areas for further research. First, as AI becomes more sophisticated and robots are deployed in more social environments, research on ethical and legal conflicts between humans and intelligent machines should be pursued vigorously. Examples of topics worthy of investigation include unmanned aerial vehicles (UAVs) in military operations, autonomous vehicles in commercial and public transportation, and medical telepresence robots. Second, the burgeoning scholarship on the ethics of environmental robots (i.e., van Wynsberghe & Donhauser, 2018; Donhauser, 2019) should assess the role of AI in environmental protection. How might robots improve our ability to address environmental challenges, and what are the practical and ethical issues associated with employing intelligent machines in such varied contexts as disaster relief, environmental justice, sustainable development, and wildlife management? Third, additional effort should be expended to understand how AI might be used to respect, protect, and fulfill *human* rights. How might robots safeguard civilians, especially vulnerable groups such as women, children, disabled people, and the elderly, during conflict? To what extent might intelligent machines enhance the efficiency and sustainability of agricultural production in order to combat food insecurity? How might artificial agents contribute to democratic governance, especially in terms of facilitating public participation in political decision-making and strengthening electoral processes? These topics represent but a few of the many fruitful avenues scholars would be wise to consider examining in the near future.

As a placard posted in Honda's "Robots in Your Life" exhibit at Japan's Miraikan (National Museum of Emerging Science and Innovation) declares, "[r]obots are already living alongside us ... what will it take for human beings to live in harmony with robots?" (Miraikan, 2019). Indeed, their pervasiveness is only likely to increase as they continue to "invade" our homes, workplaces, medical facilities, businesses, schools, and even houses of worship for reasons related to efficiency, security, health, and so on. However, "top-down" and "bottom-up" paradigms of robot design (Ishiguro, 2006, p. 321) have begun populating the world with a divergent mixture of minimally lifelike mechanoids and realistic-looking zoomorphic and anthropomorphic intelligent machines. While the growing presence of all kinds of robots in human society has elicited much scholarly debate, the latter group has generated specific attention as far as morals and laws are concerned.

Interestingly, the rise of the robots has occurred contemporaneously with a global movement to extend legal rights to nonhuman natural entities. This effort has resulted in rights being granted to animals and the environment, but in different jurisdictions, under different circumstances, and using different sources of evidence and reasoning. Perhaps due to the fairly recent acknowledgment of rights for living and non-living entities, there has thus far been little reciprocal engagement among scholars and practitioners focused on the moral or legal status of robots, animals, or nature. In writing this book, I have sought to bring these discussions into dialogue with one another towards the goal of finding a mutually intelligible answer to the question of rights for nonhumans.

In my attempt to accomplish this interdisciplinary task, I have shown how the provocative (or what some might deem heretical) issue of rights for robots is less about robots and more about the process by which we determine who (or what) deserves rights. This process has been thoroughly disrupted by the Anthropocene, which has exposed the Cartesian divide between humans and nature to be empirically inaccurate and philosophically outmoded. As such, the cast of characters eligible to make claims against other entities needs to be revised. Through the philosophical and legal analyses contained within these pages, I have charted one (but by no means the only) course for allocating rights in this new epoch. But above all, I hope that this extended intellectual exercise inspires readers to think more holistically and critically about how all entities, natural and artefactual, contribute to the collective project of worlding. Together we can create a posthuman ecology that is more inclusive, compassionate, and resilient for all.

Notes

1 Given the fact that ontological properties related to psychological and moral person-hoods feed into the qualities linked with legal personhood and there is some fuzziness as to which properties are necessary for moral personhood, I will simply use the word *personhood* when referring to the next stage in the conceptual map (unless specified otherwise).
2 But see Dennett (1996), who claims that "there is no established meaning to the word 'sentience'" (p. 66).
3 Confusingly, Schmidt and Kraemer (2006) include both the Kantian and AI-inspired versions of autonomy in their criteria for a "strong notion of autonomy," which consist of "abid[ing] by self-imposed moral laws that prove its independence from the original intentions of the creator" and "self-reflexivity," or "the capability of distancing oneself from immediate impulses by means of a self-imposed second-order system of wishes, i.e. a kind of moral law" (p. 77).

4 However, as Darling (2016) points out, popular arguments for animal rights tend to be based not on the degree to which nonhuman creatures approximate the human form, but rather the extent to which we consider them capable of human-like cognition or feeling. For a broader discussion of the philosophy and law on animal rights, see Chapter Three.

5 A search of the phrase "uncanny valley" on Google Scholar yielded over 11,000 academic works. While plenty of studies validate the uncanny valley hypothesis (i.e., Mathur & Reichling, 2016; Mathur et al., 2020), with recent scholarship even providing neurological confirmation of the effect (Pütten et al., 2019) and demonstrating its relevance in the context of zoomorphic robots (Löffler et al., 2020), some researchers have had mixed results when attempting to replicate previous efforts (i.e., Palomäki et al., 2018).

6 Although the physicality of the entity has not been discussed in great detail, Fumihide Tanaka raised the point during an interview that the tactile sensation afforded by the texture of an artificial agent's external layer goes a long way towards providing humans who interact with robots a "psychological sense of security."

7 The orientation advocated for here is not as all-encompassing as Floridi's (1999) IE, which views all morally significant entities as forms of information in an infosphere (as opposed to deep ecology's biosphere). I argue that adopting information as the primary unit of accounting in a moral universe results in a meta-ethic that is ontologically over-inclusive, producing intuitively problematic conflicts of interest and overlooking the basis of meaningful relations that occur among cognizable entities of biotic or abiotic composition.

8 AHC4806–2017 ('Chucho'), Radicación no. 17001–22–13–000–2017–00468–02, available at http://static.iris.net.co/semana/upload/documents/radicado-n-17001-22-13-000-2017-00468-02.pdf. Translated by Javier Salcedo, available at https://www.nonhumanrights.org/content/uploads/Translation-Chucho-Decision-Translation-Javier-Salcedo.pdf, at 6.

References

Airenti, G. (2015). The Cognitive Bases of Anthropomorphism: From Relatedness to Empathy. *International Journal of Social Robotics*, *7*(1), 117–127.

Arias-Maldonado, M. (2019). The "Anthropocene" in Philosophy: The Neo-material Turn and the Question of Nature. In F. Biermann & E. Lövbrand (Eds.), *Anthropocene Encounters: New Directions in Green Political Thinking* (pp. 50–66). Cambridge University Press.

Arvidsson, M. (2020). The Swarm That We Already Are: Artificially Intelligent (AI) Swarming 'Insect Drones', Targeting and International Humanitarian Law in a Posthuman Ecology. *Journal of Human Rights and the Environment*, *11*(1), 114–137.

Blaser, M. (2014). Ontology and Indigeneity: On the Political Ontology of Heterogeneous Assemblages. *Cultural Geographies*, *21*(1), 49–58.

Boyd, D. R. (2003). *Unnatural Law: Rethinking Canadian Environmental Law and Policy*. UBC Press.

Boyer, P. (1996). What Makes Anthropomorphism Natural: Intuitive Ontology and Cultural Representations. *The Journal of the Royal Anthropological Institute*, *2*(1), 83–97.

Bryson, J. J. (2018). Patiency Is Not a Virtue: The Design of Intelligent Systems and Systems of Ethics. *Ethics and Information Technology*, *20*(1), 15–26.

Calverley, D. J. (2008). Imagining a Non-Biological Machine as a Legal Person. *AI and Society*, *22*(4), 523–537.

Carruthers, P. (2005). Why the Question of Animal Consciousness Might Not Matter Very Much. *Philosophical Psychology, 18*(1), 83–102.

Castro-González, Á., Admoni, H., & Scassellati, B. (2016). Effects of Form and Motion on Judgments of Social Robots' Animacy, Likability, Trustworthiness and Unpleasantness. *International Journal of Human–Computer Studies, 90*, 27–38.

Coeckelbergh, M. (2010). Robot Rights? Towards a Social-Relational Justification of Moral Consideration. *Ethics and Information Technology, 12*(3), 209–221.

Collins, L. M. (2019). Environmental Resistance in the Anthropocene. *Oñati Socio-Legal Series*, Online, 1–19.

Cuthbertson, A. (2019, August 20). YouTube Is Deleting Videos of Robots Fighting Because of 'Animal Cruelty.' *The Independent*. Retrieved from https://www.independ ent.co.uk/life-style/gadgets-and-tech/news/youtube-robot-combat-videos-animal-cr uelty-a9071576.html.

Daly, E. (2012). *Dignity Rights: Courts, Constitutions, and the Worth of the Human Person*. University of Pennsylvania Press.

Danaher, J. (2019). Regulating Child Sex Robots: Restriction or Experimentation? *Medical Law Review, 27*(4), 553–575.

Darling, K. (2016). Extending Legal Protection to Social Robots: The Effects of Anthropomorphism, Empathy, and Violent Behavior Towards Robotic Objects. In R. Calo, A. M. Froomkin, & I. Kerr (Eds.), *Robot Law* (pp. 213–232). Edward Elgar.

De Lucia, V. (2017). Critical Environmental Law and the Double Register of the Anthropocene: A Biopolitical Reading. In L. J. Kotzé (Ed.), *Environmental Law and Governance for the Anthropocene* (pp. 97–116). Hart Publishing.

Dennett, D. C. (1976). Conditions of Personhood. In A. O. Rorty (Ed.), *The Identities of Persons* (pp. 175–196). University of California Press.

Dennett, D. C. (1996). *Kinds of Minds: Towards an Understanding of Consciousness*. Basic Books.

DiPaolo, A. (2019). If Androids Dream, Are They More Than Sheep?: Westworld, Robots and Legal Rights. *Dialogue, 6*(2). Retrieved from http://journaldialogue.org/issues/v 6-issue-2/if-androids-dream-are-they-more-than-sheep-robot-protagonists-and-human -rights/.

Donahoe, E., & Metzger, M. M. (2019). Artificial Intelligence and Human Rights. *Journal of Democracy, 30*(2), 115–126.

Donhauser, J. (2019). Environmental Robot Virtues and Ecological Justice. *Journal of Human Rights and the Environment, 10*(2), 176–192.

Duffy, B. R. (2003). Anthropomorphism and the Social Robot. *Robotics and Autonomous Systems, 42*(3), 177–190.

Epley, N., Waytz, A., & Cacioppo, J. T. (2007). On Seeing Human: A Three-Factor Theory of Anthropomorphism. *Psychological Review, 114*(4), 864–886.

Epley, N., Waytz, N., Akalis, S., & Cacioppo, J. T. (2008). When We Need a Human: Motivational Determinants of Anthropomorphism. *Social Cognition, 26*(2), 143–155.

Finnemore, M., & Sikkink, K. (1998). International Norm Dynamics and Political Change. *International Organization, 52*(4), 887–917.

Floridi, L. (1999). Information Ethics: On the Philosophical Foundation of Computer Ethics. *Ethics and Information Technology, 1*(1), 33–56.

Fowler, C. (2018). Relational Personhood Revisited. *Cambridge Archaeological Journal, 26*(3), 397–412.

Fox, W. (1990). *Toward a Transpersonal Ecology: Developing New Foundations for Environmentalism*. Shambhala.

Friedman, B. (2002). The Birth of an Academic Obsession: The History of the Countermajoritian Difficulty, Part Five. *Yale Law Journal, 112*(2), 153–260.

Fussell, S. R., Kiesler, S., Setlock, L. D., & Yew, V. (2008). How People Anthropomorphize Robots. *Proceedings of the 3rd ACM/IEEE International Conference on Human-Robot Interaction* (pp. 145–152).

Gersen, J. S. (2019). Sex Lex Machina: Intimacy and Artificial Intelligence. *Columbia Law Review, 119*(7), 1793–1809.

Goff, P. (2019). *Galileo's Error: Foundations for a New Science of Consciousness.* Pantheon Books.

Gray, K., & Wegner, D. M. (2012). Feeling Robots and Human Zombies: Mind Perception and the Uncanny Valley. *Cognition, 125*(1), 125–130.

Grear, A. (2015a). Deconstructing Anthropos: A Critical Legal Reflection on 'Anthropocentric' Law and Anthropocene 'Humanity.' *Law and Critique, 26*(3), 225–249.

Grear, A. (2015b). Towards New Legal Futures? In Search of Renewing Foundations. In A. Grear & E. Grant (Eds.), *Thought, Law, Rights and Action in the Age of Environmental Crisis* (pp. 283–313). Edward Elgar.

Griffin, D. R., & Speck, G. B. (2004). New Evidence of Animal Consciousness. *Animal Cognition, 7*(1), 5–18.

Groff, Z., & Ng, Y.-K. (2019). Does Suffering Dominate Enjoyment in the Animal Kingdom? An Update to Welfare Biology. *Biology and Philosophy, 34*(4), 40.

Gunkel, D. J. (2012). *The Machine Question: Critical Perspectives on AI, Robots, and Ethics.* MIT Press.

Haraway, D. J. (2016). *Staying with the Trouble: Making Kin in the Chthulucene.* Duke University Press.

Haslam, N. (2006). Dehumanization: An Integrative Review. *Personality and Social Psychology Review, 10*(3), 252–264.

Hill, K. (2014, July 14). Are Child Sex-Robots Inevitable? *Forbes.* Retrieved from https://www.forbes.com/sites/kashmirhill/2014/07/14/are-child-sex-robots-inevitable/.

Himma, K. E. (2009). Artificial Agency, Consciousness, and the Criteria for Moral Agency: What Properties Must an Artificial Agent Have to Be a Moral Agent? *Ethics and Information Technology, 11*(1), 19–29.

Ishiguro, H. (2006). Android Science: Conscious and Subconscious Recognition. *Connection Science, 18*(4), 319–332.

Jobin, A., Ienca, M., & Vayena, E. (2019). The Global Landscape of AI Ethics Guidelines. *Nature Machine Intelligence, 1*(9), 389–399.

Johnson, D. G. (2006). Computer Systems: Moral Entities but Not Moral Agents. *Ethics and Information Technology, 8*(4), 195–204.

Kanda, T., Miyashita, T., Osada, T., Haikawa, Y., & Ishiguro, H. (2008). Analysis of Humanoid Appearances in Human–Robot Interaction. *IEEE Transactions on Robotics, 24*(3), 725–735.

Kaplan, J. (2016). *Artificial Intelligence: What Everyone Needs to Know.* Oxford University Press.

Kateb, G. (2011). *Human Dignity.* Harvard University Press.

King, Jr., M. L. (1963, April 16). *Letter from the Birmingham Jail.* Retrieved from http://okra.stanford.edu/transcription/document_images/undecided/630416-019.pdf.

Kotzé, L. J., & Kim, R. E. (2019). Earth System Law: The Juridical Dimensions of Earth System Governance. *Earth System Governance, 1*, 1–12.

Kubes, T. (2019). New Materialist Perspectives on Sex Robots. A Feminist Dystopia/ Utopia? *Social Sciences, 8*(8), 224.

Kwak, S. S., Kim, Y., Kim, E., Shin, C., & Cho, K. (2013). What Makes People Empathize with an Emotional Robot? The Impact of Agency and Physical Embodiment on Human Empathy for a Robot. *Proceedings of the IEEE International Workshop on Robot and Human Communication (ROMAN)* (pp. 180–185).

Latour, B. (2004). *Politics of Nature*. Harvard University Press.

Lee, M. Y. K. (2010). *Equality, Dignity, and Same-Sex Marriage: A Rights Disagreement in Democratic Societies*. Martinus Nijhoff.

Lee, P., & George, R. P. (2008). The Nature and Basis of Human Dignity. *Ratio Juris, 21*(2), 173–193.

Lewis, J. E., Arista, N., Pechawis, A., & Kite, S. (2018). Making Kin with the Machines. *Journal of Design and Science, 3*, 5. doi:10.21428/bfafd97b.

Löffler, D., Dörrenbächer, J., & Hassenzahl, M. (2020). The Uncanny Valley Effect in Zoomorphic Robots: The U-Shaped Relation Between Animal Likeness and Likeability. *Proceedings of the 2020 ACM/IEEE International Conference on Human–Robot Interaction* (pp. 261–270).

Magallanes, C. I. (2015). Māori Cultural Rights in Aotearoa New Zealand: Protecting the Cosmology That Protects the Environment. *Widener Law Review, 21*(2), 273–328.

Malinowski, B. (1930). 17. Kinship. *Man, 30*, 19–29.

Malle, B. F., & Scheutz, M. (2016). Inevitable Psychological Mechanisms Triggered by Robot Appearance: Morality Included? *2016 AAAI Spring Symposium Series Technical Reports SS-16-03* (pp. 144–146).

Manzotti, R., & Jeschke, S. (2016). A Causal Foundation for Consciousness in Biological and Artificial Agents. *Cognitive Systems Research, 40*(C), 172–185.

Mathur, M. B., & Reichling, D. B. (2016). Navigating a Social World with Robot Partners: A Quantitative Cartography of the Uncanny Valley. *Cognition, 146*, 22–32.

Mathur, M. B., Reichling, D. B., Lunardini, F., Geminiani, A., Antonietti, A., Ruijten, P. A. M., Levitan, C. A., Nave, G., Manfredi, D., Bessette-Symons, B., Szuts, A., & Aczel, B. (2020). Uncanny but Not Confusing: Multisite Study of Perceptual Category Confusion in the Uncanny Valley. *Computers in Human Behavior, 103*, 21–30.

Melson, G. F., Kahn, Jr., P. H., Beck, A., & Friedman, B. (2009). Robotic Pets in Human Lives: Implications for the Human–Animal Bond and for Human Relationships with Personified Technologies. *Journal of Social Issues, 65*(3), 545–567.

Miraikan (2019). *Robots in Your Life* [Museum label]. Tokyo, Japan: Author.

Miralles, A., Raymond, M., & Lecointre, G. (2019). Empathy and Compassion Toward Other Species Decrease with Evolutionary Divergence Time. *Scientific Reports, 9*(1), 1–8.

Mithen, S. (1996). *The Prehistory of the Mind: The Cognitive Origins of Art, Religion and Science*. Thames and Hudson.

Mori, M. (1970). The Uncanny Valley. *Energy, 7*, 33–35.

Naess, A. (1973). The Shallow and the Deep, Long-Range Ecology Movement: A Summary. *Inquiry, 16*(1–4), 95–100.

Nagel, T. (1979). Panpsychism. In T. Nagel (Ed.), *Mortal Questions*. Cambridge University Press.

Neave, C. E. (1909). A Friend to Dumb Animals. *The English Illustrated Magazine, 72*, 563–566.

Nucci, E. D. (2017). Sex Robots and the Rights of the Disabled. In J. Danaher & N. McArthur (Eds.), *Robot Sex: Social and Ethical Implications* (pp. 73–88). MIT Press.

Nyholm, S., & Frank, L. E. (2019). It Loves Me, It Loves Me Not: Is It Morally Problematic to Design Sex Robots That Appear to Love Their Owners? *Techne: Research in Philosophy and Technology, 23*(3), 402–424.

Palomäki, J., Kunnari, A., Drosinou, M., Koverola, M., Lehtonen, N., Halonen, J., Repo, M., & Laakasuo, M. (2018). Evaluating the Replicability of the Uncanny Valley Effect. *Heliyon*, *4*(11), e00939.

Pelizzon, A., & Gagliano, M. (2015). The Sentience of Plants: Animal Rights and Rights of Nature Intersecting? *Australian Animal Protection Law Journal*, *11*, 5–14.

Philippopoulos-Mihalopoulos, A. (2013). Actors or Spectators? Vulnerability and Critical Environmental Law. *Oñati Socio-Legal Series*, *3*(5), 854–876.

Philippopoulos-Mihalopoulos, A. (2017). Critical Environmental Law in the Anthropocene. In L. J. Kotzé (Ed.), *Environmental Law and Governance for the Anthropocene* (pp. 117–136). Hart Publishing.

Posthumus, D. C. (2017). All My Relatives: Exploring Nineteenth-Century Lakota Ontology and Belief. *Ethnohistory*, *64*(3), 379–400.

Pütten, A. M. R. der, Krämer, N. C., Maderwald, S., Brand, M., & Grabenhorst, F. (2019). Neural Mechanisms for Accepting and Rejecting Artificial Social Partners in the Uncanny Valley. *Journal of Neuroscience*, *39*(33), 6555–6570.

Richardson, K. (2016). Sex Robot Matters: Slavery, the Prostituted, and the Rights of Machines. *IEEE Technology and Society Magazine*, *35*(2), 46–53.

Risse, M. (2019). Human Rights and Artificial Intelligence: An Urgently Needed Agenda. *Human Rights Quarterly*, *41*(1), 1–16.

Sætra, H. S. (2020). The Parasitic Nature of Social AI: Sharing Minds with the Mindless. *Integrative Psychological and Behavioral Science*. doi:10.1007/s12124-020-09523-6.

Sahlins, M. (2011). What Kinship Is (Part One). *Journal of the Royal Anthropological Institute*, *17*(1), 2–19.

Salmón, E. (2000). Kincentric Ecology: Indigenous Perceptions of the Human–Nature Relationship. *Ecological Applications*, *10*(5), 1327–1332.

Schmidt, C. T. A., & Kraemer, F. (2006). Robots, Dennett and the Autonomous: A Terminological Investigation. *Minds and Machines*, *16*(1), 73–80.

Schneider, S. (2016, March 18). The Problem of AI Consciousness. *Huffington Post*. Retrieved from https://www.huffpost.com/entry/the-problem-of-ai-conscio_b_9502790.

Searle, J. R. (1980). Minds, Brains, and Programs. *Behavioral and Brain Sciences*, *3*(3), 417–457.

Searle, J. R. (2008). Animal Minds. In J. Feinberg & R. Shafer-Landau (Eds.), *Reason and Responsibility: Readings in Some Basic Problems of Philosophy* (13th ed., pp. 356–365). Thomson: Wadsworth.

Shelton, D. L. (2014). *Advanced Introduction to International Human Rights Law*. Edward Elgar.

Shevlin, H., Vold, K., Crosby, M., & Halina, M. (2019). The Limits of Machine Intelligence. *EMBO Reports*, *20*(10), e49177.

Singer, P. (1974). All Animals Are Equal. *Philosophic Exchange*, *5*(1), 103–116.

Solum, L. B. (1992). Legal Personhood for Artificial Intelligences. *North Carolina Law Review*, *70*(4), 1231–1288.

Sparrow, R. (2002). The March of the Robot Dogs. *Ethics and Information Technology*, *4*(4), 305–318.

Sparrow, R. (2017). Robots, Rape, and Representation. *International Journal of Social Robotics*, *9*(4), 465–477.

Sternberg, R. J. (2000). The Concept of Intelligence. In R. J. Sternberg (Ed.), *Handbook of Intelligence* (pp. 3–15). Cambridge University Press.

Studley, J. (2019). *Indigenous Sacred Natural Sites and Spiritual Governance: The Legal Case for Juristic Personhood*. Routledge.

Sullins, J. P. (2006). When Is a Robot a Moral Agent? *International Review of Information Ethics*, *6*, 23–30.

Tasioulas, J. (2019). First Steps Towards an Ethics of Robots and Artificial Intelligence. *Journal of Practical Ethics*, *7*(1), 61–95.

Teubner, G. (2006). Rights of Non-Humans? Electronic Agents and Animals as New Actors in Politics and Law. *Journal of Law and in Society*, *33*(4), 497–521.

Torrance, S. (2008). Ethics and Consciousness in Artificial Agents. *AI and Society*, *22*(4), 495–521.

Turing, A. M. (1950). I.—Computing Machinery and Intelligence. *Mind*, *59*(236), 433–461.

Turner, J. (2019). *Robot Rules: Regulating Artificial Intelligence*. Palgrave Macmillan.

UN General Assembly (1948). *Universal Declaration of Human Rights* (217 [III] A). Paris. Retrieved from https://www.un.org/en/universal-declaration-human-rights/.

Vallverdú, J. (2011). The Eastern Construction of the Artificial Mind. *Enrahonar: Quaderns de Filosofia*, *47*, 171–185.

van Wynsberghe, A., & Donhauser, J. (2018). The Dawning of the Ethics of Environmental Robots. *Science and Engineering Ethics*, *24*(6), 1777–1800.

Vogel, S. (2015). *Thinking Like a Mall: Environmental Philosophy After the End of Nature*. MIT Press.

Wagman, M. (1999). *The Human Mind According to Artificial Intelligence: Theory, Research, and Implications*. Praeger.

Warren, M. A. (1997). *Moral Status: Obligations to Persons and Other Living Things*. Oxford University Press.

Watson, R. A. (1979). Self-Consciousness and the Rights of Nonhuman Animals and Nature. *Environmental Ethics*, *1*(2), 99–129.

Waytz, A., Cacioppo, J. T., & Epley, N. (2010). Who Sees Human?: The Stability and Importance of Individual Differences in Anthropomorphism. *Perspectives on Psychological Science*, *5*(3), 219–232.

Winfield, A. (2014, August 9). Artificial Intelligence Will Not Turn Into a Frankenstein's Monster. *The Guardian*. Retrieved from https://www.theguardian.com/technology/2014/aug/10/artificial-intelligence-will-not-become-a-frankensteins-monster-ian-winfield.

Wise, S. M. (2002). *Drawing the Line: Science and the Case for Animal Rights*. Perseus Books.

Wise, S. M. (2013). Nonhuman Rights to Personhood. *Pace Environmental Law Review*, *30*(3), 1278–1290.

Wohlleben, P. (2016). *The Hidden Life of Trees: What They Feel, How They Communicate— Discoveries from A Secret World* (J. Billinghurst, Trans.). Greystone Books.

Young, J. E., Sung, J., & Voida, A. (2011). Evaluating Human–Robot Interaction: Focusing on the Holistic Interaction Experience. *International Journal of Social Robotics*, *3*(1), 53–67.

Zardiashvili, L., & Fosch-Villaronga, E. (2020). "Oh, Dignity too?" Said the Robot: Human Dignity as the Basis for the Governance of Robotics. *Minds and Machines*. doi:10.1007/s11023-019-09514-6.

Złotowski, J. A., Strasser, E., & Bartneck, C. (2014). Dimensions of Anthropomorphism: From Humanness to Humanlikeness. *Proceedings of the 2014 ACM/IEEE International Conference on Human–Robot Interaction* (pp. 66–73).

Złotowski, J. A., Sumioka, H., Nishio, S., Glas, D. F., Bartneck, C., & Ishiguro, H. (2015). Persistence of the Uncanny Valley: The Influence of Repeated Interactions and a Robot's Attitude on its Perception. *Frontiers in Psychology*, *6*, 883.

Index

Page numbers in bold indicate a table.

Printed in the United States
By Bookmasters